# CURRENT TOPICS IN LACTATION

Edited by **Isabel Gigli**

**Current Topics in Lactation**
http://dx.doi.org/10.5772/63182
Edited by Isabel Gigli

**Contributors**

Edyta Molik, Eveline Ibeagha-Awemu, Duy Do, Edith Kordon, Omar Coso, Julia Halperin, Veronica Berta Dorfman, Alfredo Daniel Vitulo, Isabel Gigli

**Published by InTech**
Janeza Trdine 9, 51000 Rijeka, Croatia



**Notice**

Statements and opinions expressed in the chapters are these of the individual contributors and not necessarily those of the editors or publisher. No responsibility is accepted for the accuracy of information contained in the published chapters. The publisher assumes no responsibility for any damage or injury to persons or property arising out of the use of any materials, instructions, methods or ideas contained in the book.

**Publishing Process Manager** Dajana Pemac
**Technical Editor** SPi Global
**Cover** InTech Design team

First published May, 2017
Printed in Croatia
Legal deposit, Croatia: National and University Library in Zagreb

Additional hard copies can be obtained from orders@intechopen.com

Current Topics in Lactation, Edited by Isabel Gigli
p. cm.
Print ISBN 978-953-51-3137-3
Online ISBN 978-953-51-3138-0

For EU product safety concerns:
IN TECH d.o.o., Prolaz Marije Krucifikse Kozulić 3, 51000 Rijeka, Croatia,
info@intechopen.com or visit our website at intechopen.com.

# Meet the editor

Isabel Gigli, graduated Veterinary Medicine and a Doctor of Science, has conducted her doctoral research training at the Institute of Biology and Experimental Medicine (IByMe), in Buenos Aires, Argentina. Afterward, she completed a three-year postdoctoral training at the University of Cornell, USA. In addition, she worked for two years at the University of Palermo, Italy. Currently, Dr. Gigli is a full-time professor at the School of Agriculture at the National University of La Pampa (UNLPam), Argentina, where she teaches physiology of lactation. Dr. Gigli has published several scientific articles and two books. She also edited the book Milk Proteins - From Structure to Biological Properties and Health Aspects published by InTech.

Contents

# Preface

In recent decades, the development of molecular techniques such as deep sequencing and RNA sequencing contributes to increase significantly our knowledge of molecular mechanisms in lactation. In this book, readers will find updated research on mammary gland development, endocrine regulation of lactation, and mammary gland involution. Although the physiological bases on milk secretion are fairly consistent among species, some mechanisms may be different. In this regard, comparative research increases our basic knowledge and helps us to understand processes that may be less evident in some species. From studies in cows, sheep and mice, and less studied rodents such as vizcacha, each chapter of the present book provides a full review on important topics in lactation.

This book is an excellent opportunity for graduate students and researchers to read updated information on this field. *Current Research in Lactation* will be a valuable resource for dairy physiologists and biomedical scientists interested in mammary glands and lactation.

**Isabel Gigli**
School of Agriculture, National University of La Pampa (UNLPam)
Santa Rosa, La Pampa
Argentina

Chapter 1

# Introductory Chapter: Insights into Lactation

Isabel Gigli

Additional information is available at the end of the chapter

http://dx.doi.org/10.5772/intechopen.68395

Lactation is a fascinating process that characterized all mammals. From primitive glands with no nipples—as in monotremes—to complex mammary glands, female mammals nourish their offspring with the perfect food. Each species has different nutritional requirements, and each mother secretes the correct concentration of macro and micronutrients. Fat, for instance, varies from 50% content as in gray seal [1] to 4% in cow milk [2]. In addition to nutrition, milk contains bioactive components that promote immunity such as immunoglobulin and lactoferrin [3, 4]. The importance of milk on the prevention of pathogens in the newborn has been known for a long time [5], yet as research progresses, new understanding arises. Recently, it was identified that human milk produces a large amount of oligosaccharides that are not digestible by the newborn. Researchers found that these short carbohydrates play an important role on the intestinal flora acting as prebiotic for beneficial bacteria and inhibiting the adhesion of pathogens on the intestinal epithelial [6–8]. Human milk is not the only one that produces prebiotic components. In cows, also have been identified a variety of oligosaccharides that promote a healthy microbiome [9]. All these recent studies highlight even more the impact of milk as a functional food.

The mammary gland develops through different phases: embryonic, pre- and post-puberal, and during pregnancy. An important feature in the mammary gland is that it undergoes different cycles of differentiation and regression throughout the adult female's life. Lactation is the final stage of reproduction; therefore, lactation regulation might differ according to the species' reproduction strategy. Female mammals can be classified as continuous or seasonal breeders. Concomitantly, seasonal breeder species can be divided into long and short day. The females on the first group present ovary activity in spring and summer, while are in anestrous in fall and winter (i.e., horses). On the contrary, in short-day breeders, reproduction takes place in fall and winter and ovary is inactive in spring and summer (i.e., sheep). In all seasonal breeders, regarding the reproduction strategy, the decreased light length produces the excitation of the superior cervical ganglion through the retina nerve, which unblocks the inhibition of the pineal gland. As a result, melatonin increases. Melatonin interacts with the gonadotropin-release hormone (GnRH) affecting the hypothalamus pineal ovary axis. The way melatonin influences GnRH is depending on the reproduction strategy: in short-day breeders, melatonin induces the synthesis of GnRH, and therefore, ovary cycles are produced; on the other hand, in long-day breeder, melatonin inhibits GnRH-inducing anestrous [10]. Sheep—which are short-day breeders—can be artificially manipulated to induce cycling out of season. Melatonin also affects mammary gland and lactation. Therefore, lactation in sheep lambing out of season could affect milk production. In Chapter 1, Dr. Molik describes the endocrine response associated with melatonin changes and the impact of the photoperiod on milk production in the sheep.

Another example of how reproduction strategy influences mammary gland development is observed in *Lagostomus maximus*. This rodent presents a pseudo-ovulation in mid-gestation. The changes in progesterone levels during pregnancy characterized the mammary gland development of this species. An interesting description of the mammary gland development and involution on the Vizcacha is described in Chapter 2 by Dr. Halperin et al.

Numerous factors interact at the onset and maintaining of lactation. There is a consensus of the main lactogenic hormones that regulate milk synthesis and secretion, but as research advances, we are getting a better understanding on the fine balance among the systemic hormones and the local regulatory factors. Milk secretion is trigger at parturition when progesterone falls and glucocorticoid, prolactin, and growth hormone (GH) rise [4]. The association among these hormones and lactation has been known for many years. Probably, one of the earlier evidence that progesterone inhibits lactation was back in 1925, when Hammond described that milk secretion in rabbits was the consequence of corpora lutea involution [11]. Prolactin is needed to maintain milk secretion in most mammals. However, in ruminants, the predominant lactogenic hormone is considered to be GH [12]. A synthetic GH was widely used in dairy industry to increase milk yield in the 1980s when a recombinant GH was commercially available [13]. At a molecular level, lactogenic hormones activate gene expression through JAK2/STAT5. Briefly, once the hormones bind the cell surface receptors, JAK2 proteins phosphorylate and consequently induce the phosphorylation of STAT5. Then, the phosphorylated dimer STAT5 translocates to the nucleus and induces transcription of target milk protein genes. For a review, see Ref. [14]. Milk is synthesized and secreted into the mammary alveoli lumen. When milk let-down stimulus is triggered, oxytocin is released from the pituitary gland and binds to its receptor on the myoepithelial cells; as a consequence, milk is ejected. Oxytocin structure and function have been described in the 1950s. Vincent du Vigneaud received a Nobel Prized in 1955 for his work on the oxytocin [15]. Maintaining of lactation requires the periodic removal of milk. Systemic endocrine factors and autocrine mechanisms act concomitantly to control mammary gland function. Once milking is discontinuous, the mammary gland undergoes to a series of involution process that drive it from a secretory phase to an inactive organ. Involution is divided into two stages. The first stage is regulated by local factors, which outweigh the positives stimulus exerted by the lactogenic hormones. To exemplify the importance of this concept, interrupting milking in a single gland in the cow, triggers the involution process only on that quarter, while the other three lands maintain milk production. During the second involution stage, apoptosis of epithelial cells and tissue remodeling occurs. In Chapter 3 of the present book, Dr. Kordon and Dr. Coso describe in detail the cell signaling associated with mammary gland involution and cancer.

Authors explain the role of STAT3 and leukemia inhibitory factor (LIF) on mammary cell death and show that the effect of stretching mammary epithelial cells in culture—as happen in vivo by milk accumulation—induces both LIF expression and STAT3 phosphorylation. Also, it discussed the role of the extracellular matrix in the involution process.

Local mechanisms not only regulate milk yield but also control milk composition. A beautiful example of local regulation in milk composition is observed in the Tammar Wallaby. When the immature joey is born, it crawls to the mother's pouch and attaches to the nipple. As the youth growths and its nutritional requirement change, milk composition is adapted to the new demand. However, it could occur that the female gives birth to a second joey, while the oldest one is still lactating. On this occasion, entirely different composition of milk would be produced in each mammary gland. This is called asynchronous concurrent lactation and is regulated by local factors. For an extensive review, see Ref. [16].

The development of new molecular tools—such as new-generation sequencing and RNA sequencing—revealed remarkable information that challenges our knowledge. In 1998, Fire and Mello received a Nobel Prize for their discovery of small RNA (miRNA) that control proteins synthesis post-transcriptional in Caenorhabditis [17]. By 2001, it was shown that these miRNAs far from being specific in worms were small molecules that regulate gene expression in eukaryote. Since then, thousands of miRNAs have been identified in different organs and specifically in the mammary gland and even in milk secretion. In mammary gland, miRNAs modulate development and regression. Several miRNAs are secreted in milk inside vesicles that give them protection from the low gastric pH. It was postulated that milk miRNAs could regulate gene expression in the newborn. Still it is controversial, and further research is needed to indicate if milk miRNAs are biological active ones there are absorbed [18, 19]. In Chapter 4, Dr. Duy and Dr. Ibeagha-Awemu describe in detail the state-of-the-art on noncoding RNA in the ruminant mammary gland.

Our knowledge on mammary gland physiology has increased considerably in the last decades. This book offers to the readers an update on research in four important areas: influence of melatonin in lactation, mammary gland development, signaling process in the mammary gland involution, and the role of miRNAs in mammary gland physiology. The study of comparative physiology lactation contributes to understanding biological process than could be present in all species but less evident in some and provides a wider understanding of the lactation process. The discovery of miRNAs definitely opens a new era in the study of gene expression. All these studies—plus the new molecular technology available—have increased and certainly will increase even more, our knowledge in the mammary gland and lactation, a captivating physiology process.

## Author details

Isabel Gigli

Address all correspondence to: igigli@agro.unlpam.edu.ar

School of Agriculture - National University of La Pampa (UNLPam), Santa Rosa, La Pampa, Argentina

## References

[1] Baker JR. Grey seal (*Halichoerus grypus*) milk composition and its variation over lactation. British Veterinary Journal. 1990;**146**(3):233–238

[2] Akers RM. Functional development of the mammary gland. In: Lactation and the Mammary Gland. Chapter 3. Wiley-Blackwell; Hoboken, New Jersey, United States, 2002. pp. 66–87. ISBN: 978-0-8138-2992-0

[3] Ribeiro ST, Vainstein MH, Ulhoa CJ. Free secretory component and lactoferrin of human milk inhibit the adhesion of enterotoxigenic *Escherichia coli*. Journal of Medical Microbiology. 1995;**42**(1):3–9

[4] Neville MC. Lactation and its hormonal control. Chapter 57. In: Physiology of Reproduction. 3rd ed. Knobil and Neill's. Academic Pres, Elsevier; Cambridge, Massachusetts, United States; 2006. pp. 2983-3036. eBook ISBN: 9780080535272

[5] Mccarthy EF, Mcdougall EI. Absorption of immune globulin by the young lamb after ingestion of colostrum. Nature. 1949;**164**(4165):354

[6] Manthey CF, Autran CA, Eckmann L, Bode LJ. Human milk oligosaccharides protect against enteropathogenic Escherichia coli attachment in vitro and EPEC colonization in suckling mice. J Pediatr Gastroenterol Nutr. 2014;**58**(2):165–168

[7] Boudry G, Hamilton MK, Chichlowski M, Wickramasinghe S, Barile D, Kalanetra KM, Mills DA, Raybould HE. Bovine milk oligosaccharides decrease gut permeability and improve inflammation and microbial dysbiosis in diet-induced obese mice. Journal of Dairy Science. 2017;**100**:2471–2481. pii: S0022-0302(17)30072-3. DOI: 10.3168/jds.2016-11890

[8] Lammerts van Bueren A, Mulder M, Leeuwen SV, Dijkhuizen L. Prebiotic galactooligosaccharides activate mucin and pectic galactan utilization pathways in the human gut symbiont *Bacteroides thetaiotaomicron*. Scientific Reports. 2017;**7**:40478.

[9] Mehra R, Barile D, Marotta M, Lebrilla CB, Chu C, German JB. Novel high-molecular weight fucosylated milk oligosaccharides identified in dairy streams. PLoS One. 2014;**9**(5):e96040

[10] Ginther OJ. In Reproductive Biology of the Mare: Basic and Applied Aspects published by Equiservices; 1981. 413p. DOI: http://dx.doi.org/10.1016/S0737-0806(81)80003-2

[11] Hammond J and Marshall FHA. Reproduction in the Rabbit. Edinburgh: Oliver and Boyd; 1925. 210p

[12] Peel CJ, Fronk TJ, Bauman DE, Gorewit RC. Effect of exogenous growth hormone in early and late lactation on lactational performance of dairy cows. Journal of Dairy Science. 1983;**66**(4):776–782

[13] Bauman DE, Eppard PJ, DeGeeter MJ, Lanza GM. Responses of high-producing dairy cows to long-term treatment with pituitary somatotropin and recombinant somatotropin. Journal of Dairy Science. 1985;**68**(6):1352–1362

[14] Watson CJ, Burdon TG. Prolactin signal transduction mechanisms in the mammary gland: The role of the Jak/Stat pathway. Reviews of Reproduction. 1996;**1**:1–5. Review

[15] du Vigneaud V. Oxytocin, the principal oxytocic hormone of the posterior pituitary gland: Its isolation, structure, and synthesis. Experientia. 1955;**10**(Suppl 2):9–26

[16] Brennan AJ, Sharp JA, Lefevre C, Topcic D, Auguste A, Digby M, Nicholas KR. The tammar wallaby and fur seal: Models to examine local control of lactation. Journal of Dairy Science. 2007;**90**(Suppl 1):E66–E75

[17] Fire A, Xu S, Montgomery MK, Kostas SA, Driver SE, Mello CC. Potent and specific genetic interference by double-stranded RNA in *Caenorhabditis elegans*. Nature. 1998; **391**(6669):806–811

[18] Melnik BC, Kakulas F, Geddes DT, Hartmann PE, John SM, Carrera-Bastos P, Cordain L, Schmitz G. Milk miRNAs: Simple nutrients or systemic functional regulators?. Nutrition & Metabolism (London). 2016;**13**:42

[19] Auerbach A, Vyas G, Li A, Halushka M, Witwer K. Uptake of dietary milk miRNAs by adult humans: A validation study. F1000Research. 2016;**721**:1–14. DOI: 10.12688/f1000research.8548.1

Chapter 2

# Ovarian, Hypophyseal and Hypothalamic Hormones Coordinate Mammary Gland Remodeling in Adult *Lagostomus maximus*: a Rodent that Shows Pseudo-Ovulation at Mid-Gestation

Julia Halperin, Veronica B. Dorfman and Alfredo D. Vitullo

Additional information is available at the end of the chapter

http://dx.doi.org/10.5772/66155

**Abstract**

Adult female mammary glands go through extensive tissue remodeling during pregnancy, lactation and after the weaning of the neonates. Here we characterize mammary gland morphology of adult females of *Lagostomus maximus*, a hystricomorph rodent with a pseudo-ovulatory event at mid-gestation, and describe how the glandular tissue changes its architecture in response to variations of the hormonal environment. At mid-gestation, pseudo-ovulation is seen as an essential event increasing the number of secondary corpora lutea and thus rising the circulating levels of progesterone that help to maintain pregnancy to term. As a side effect, mammary gland development is favored early during the long-lasting pregnancy of *L. maximus*, preparing females for the nutritional need of fully developed pups in this k-strategist species.

**Keywords:** mammary gland, prolactin, estradiol, progesterone, *Lagostomus maximus*

## 1. Introduction

Lactation has evolved as a vital part of the mammalian reproduction strategy [1]. During this process, ovarian, hypophyseal and hypothalamic hormones together with a myriad of factors synchronize actions for the growing and remodeling of the mammary glands. Over the past years, our understanding on how this complex hormone-driven process coordinate mechanisms to guide mammary glands throughout growth-lactation-regression cycles has greatly improved. Nonetheless, there is still much to be learned about their roles in the development of each of the structural components of this organ.

The vast majority of mammary glands investigations have been performed in mice and rats. However, many aspects still remain unfulfilled covered by these conventional animal models since they differ considerably in mammary glands development and types of breast cancer from women [2, 3]. On the other hand, studies performed on unconventional rodents such as guinea pigs and hamsters that share with humans some endocrine and reproductive biology aspects have contributed to a better understanding of human physiology and disease [4], particularly on some reproductive tumors [5, 6].

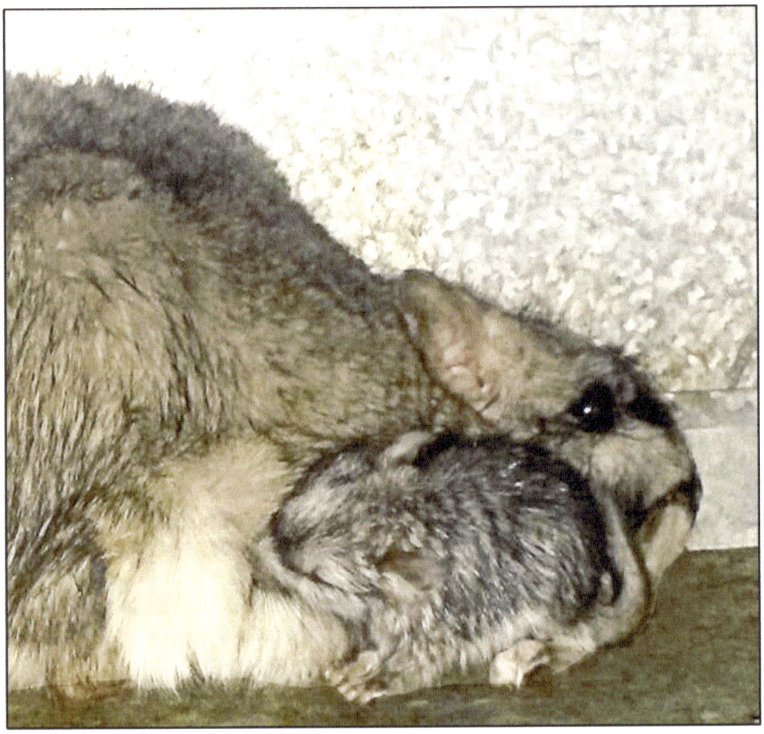

**Figure 1.** Adult female plains vizcacha (*Lagostomus maximus*) nursing a pup. Credit: J. Halperin, Universidad Maimónides.

The South American plains vizcacha, *Lagostomus maximus*, is a hystricomorph rodent closely related to guinea pig (**Figure 1**) [7]. This species has attracted significant attention in the reproductive research field since female ovaries exhibit exceptional and unique characteristics among rodents. Females display natural massive poly-ovulation that can go up to 800 oocytes per cycle, the highest ovulatory rate so far recorded for a mammal, as a result of an unusual constitutive suppression of apoptosis that greatly decreases intra-ovarian oocyte dismissal caused by follicular atresia [8–12]. In addition, gestation lasts 154 ± 6 days [8], an unusually long period for a rodent and one of the longest recorded among hystricomorphs. Moreover, pregnant females exhibit an ovulatory event at mid-pregnancy that leads to a considerable number of secondary corpora lutea with oocyte retention (i.e., pseudo-ovulation) and to an important rise of the progesterone levels [10, 11, 13]. This boost up in the circulating progesterone may contribute to an accurate maintenance of the uterus and embryo development up to the end of pregnancy [11, 14]. Given that ovarian hormones modulate growth and development of post-pubertal mammary glands, the reproductive peculiarities of the ovaries of *L. maximus* make of this species an interesting model to examine the mammary glands morphology according to its reproductive status.

The purpose of this chapter is to give a brief representation of the morphological changes that occur in the mammary glands between pregnancies under the action of ovarian, hypophyseal and hypothalamic hormones of adult plains vizcachas.

## 2. The mammary gland morphology of vizcachas

Adult female vizcachas have two pairs of functional mammary glands located below the ventral skin and laterally on the thorax. The skin epidermis is formed by a stratified squamous and keratinized epithelium which rests on a layer of dense collagenous connective tissue that contains hair follicles, sweat glands and fibroblasts. From the opening of the nipple and into the mammary glands, the number of epithelial layers decreases until it reaches a two-layer epithelium which upholsters each branched tubulo-alveolar gland [15].

The mammary gland secretory parenchyma is divided into lobes and then lobules by connective tissue septa. Lobules are formed by intralobular ducts that connect to an interlobular duct which finally empty into the lactiferous duct. The lactiferous duct is the excretory duct of each lobe and connects to the opening nipple to allow the release of milk during lactation. Before reaching the opening nipple, the lactiferous duct lumen forms a lactiferous sinus that functions as a reservoir for milk during lactation.

The mammary gland epithelium that coats the ducts is composed by an inner layer of secretory cells and an outer layer of myoepithelial cells which lies on the basement membrane that separates parenchymal and stromal compartments. The surrounding stroma is mainly composed by connective tissue, endothelial vessels, fibroblasts and immune cells. Unlike to what have been described for mouse and rat, mammary glands of adult vizcachas have a poor fat content [15]. General morphology and a detailed description of each cellular component of adult mammary glands of *L. maximus* are depicted in **Figure 2**.

As adult females transit throughout pregnancy and lactation, the mammary gland develops a more elaborated structure as a result of proliferation, branching and differentiation of the ductal tree. The extent of the development of the ductal network is closely related to the female reproductive status and the hormonal milieu.

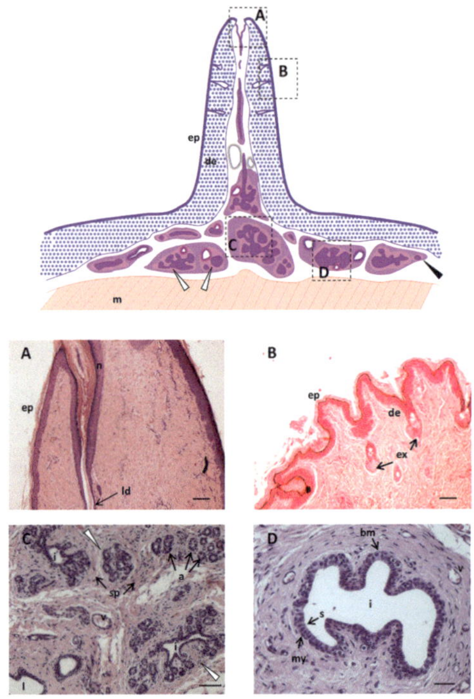


**Figure 2.** General morphology of the mammary gland of female *L. maximus*. Up: schematic draw depicts the general morphology of a post-pubertal mammary gland of *L. maximus*. Bottom: representative photomicrographs of each cellular component of the mammary glands. (A) Lactiferous duct epithelium is continuous with the stratified squamous and keratinized skin epidermis. (B) Underlying dermis is formed by a layer of dense collagenous connective tissue containing hair follicles, sweat glands and fibroblasts. (C) Secretory parenchyma composed by intralobular ducts forming lobules that join into lobes. Connective tissue septa surround each lobule. (D) Epithelial lining of the ductal network is made up of a luminal secretory cell layer and a basal myoepithelial cell layer which rests on a basal membrane that separates the epithelium from the surrounding stroma. a, secretory alveolus; bm, basal membrane; de, dermis; ep, epidermis; ex, exocrine gland; i, intralobular duct; I, interlobular duct; m, muscle; my, myoepithelial cell layer; n, nipple opening; s, secretory cell layer; sp, connective tissue septum; v, blood vessel. Filled arrowhead: lobes. Empty arrowhead: lobules. Scale bar is 150 μm for photos A and B; 100 μm for photo C and 25 μm for photo D.

## 3. Hormonal regulation of mammary glands growth and development according to the vizcacha reproductive status

### 3.1. Cycling

Short before the breeding season, mammary glands of non-pregnant adult vizcachas are in a "resting" state and present predominance of stromal connective tissue over the rudimentary ductal tree, which is mainly characterized by a few ducts and scarce secretory alveoli. At this stage, circulating estradiol can be high if the animal is ovulating. Yet, expression of estrogen receptor β, ERβ, is weak, and ERα is almost absent in the mammary glands (**Figure 3**). These observations could be interpreted as an indication that estradiol does not play an important role in the metabolism of cycling mammary glands of *L. maximus*. Such hypothesis is opposed to what have been previously shown in mammary glands of virgin mice [16–18]. Those reports demonstrated that estradiol is a crucial regulator of branching through both of its receptors, being ER$\alpha$ the more important for mammary glands development. Considering that our group of cycling vizcachas is composed by adult females captured in their natural environment, they most likely have gone through one or more pregnancies prior to the capture. Their mammary glands have already experienced pregnancy-lactation-regression cycles. Moreover, although at this resting state there is no secretory activity in the mammary glands, some ducts still show residual milk fat globules in their lumen which is indicative of a recent lactation. These evidences support the idea that these are not virgin females and so, their mammary glands are already mature. Their ductal network, even in a resting status, already comprises secondary branching. Nevertheless, the normal expression of ER$\alpha$ and ER$\beta$ in mammary gland of virgin vizcachas is still pending. Just then, we will be able to confirm the role of those receptors in the regulation of mammary gland secondary branching.

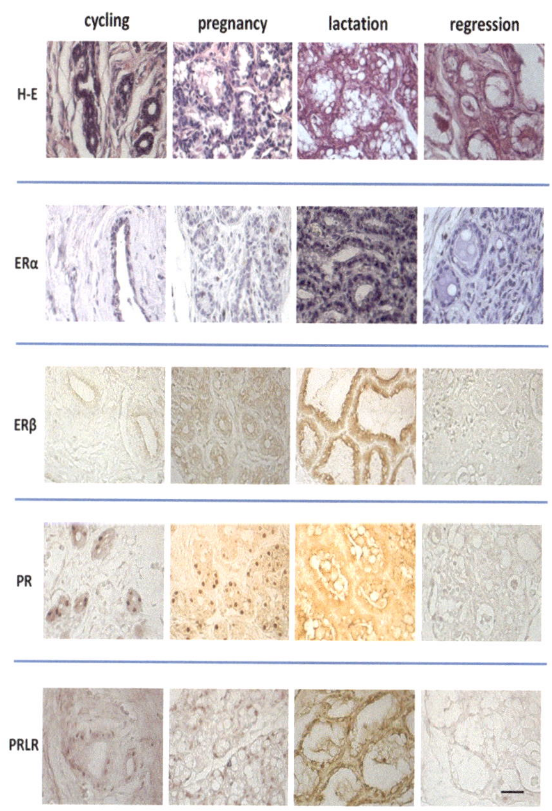


**Figure 3.** Hormonal regulation of the vizcacha mammary gland development according to the reproductive status. Representative photomicrographs of mammary gland sections of adult vizcachas at cycling, pregnancy, lactation and regression status. H-E, hematoxilyn-eosin; ER$\alpha$, estrogen receptor $\alpha$ ER$\beta$, estrogen receptor $\beta$ PR, progesterone receptor; PRLR, prolactin receptor. Immunoreactivity is shown in brown and only for ER$\alpha$ hematoxylin-counterstained nuclei in blue. All photomicrographs have the same magnification. Scale bar is depicted in the last photo (bottom right) and represents 25 μm.

It is well established that in response to ovarian steroids at the onset of cyclicity, the mammary gland enlarges, the ducts undergo rapid extension and branching, and the mammary epithelial cells fill the mammary fat pad. It is also known that, in cycling females, prolactin (PRL) is only indirectly involved in the formation of ductal side branching by promoting luteal progesterone synthesis, as evident by the restoration of ductal branching in PRL knockout females treated with progesterone [19, 20]. In accordance with these references, we did not detected PRL receptor (PRLR) expression in membrane of ductal epithelium of cycling vizcachas (**Figure 3**). Yet, we detected a conspicuous PRLR mark in nuclei if ductal epithelium. It has been proposed that polypeptide ligands like PRL and their receptors may translocate into the nucleus and regulate the expression of specific transcription factors [21]. Our results suggest that the role of PRL over mammary glands may not be restricted to its known trophic effect during pregnant and lactation phases, but it also could be modulating other physiological processes in mammary glands of non-pregnant animals. In fact, it has been shown that intact transmembrane PRLR localizes in the nucleus of human breast carcinoma cells where it functions as a co-activator through interaction with the latent transcription factor Stat5a and the high mobility group N2 protein (HMGN2) and contributes to the expression of the ER and progesterone receptor (PR) [22, 23].

### 3.2. Pregnancy

During this stage, mammary glands have to undergo further development and morphological changes in preparation for nutrition of neonates. It has been already established that progesterone induces extensive side-branching and alveologenesis and, in combination with PRL, promotes the differentiation of the alveoli, which are the structures that synthesize and secrete milk during lactation [24].

Along pregnancy, mammary glands of vizcachas increase the parenchymal-stromal ratio as well as the vascularization that surrounds each lobule. We observed that, during the first half of pregnancy of *L. maximus*, there is an increase in branching and elongation of the ductal tree accompanied by an increased expression of PRLR and of progesterone receptor (PR) expression in nuclei of secretory epithelium (**Figure 3**). Bulbous terminal end buds (TBEs) formed at the tip of growing ducts during ductal morphogenesis, now proliferate and bifurcate generating new branches. TEBs show multiple layers of epithelium implying a high proliferative rate of this cell population during gestation. Particularly, after pseudo-ovulation takes place, mammary ductal network becomes noticeably more ramified: the alveolar buds located at the end of the branches progressively cleave and differentiate into individual alveoli which occupy the majority of the fat pad (**Figure 3**) [15].

*L. maximus* shows two well defined phases during pregnancy: before and after pseudo-ovulation. In the first half of pregnancy, around day 70 of gestation, circulating progesterone gradually decreases as a result of normal luteolysis. Approximately at day 90, when circulating progesterone reaches its minimum level, a new wave of follicular recruitment, pseudo-ovulation and luteinization occurs and the released luteal progesterone progressively increases its levels throughout the second half of gestation [25].

http://dx.doi.org/10.5772/66155

Considering that progesterone is known as a key factor in the regulation of post-pubertal mammary gland development, it is interesting to note that although its levels drastically change throughout pregnancy of *L. maximus*, the pattern of PR immunoreactivity in the secretory alveoli of mammary glands remains relatively constant (**Figure 3**). The enhancement of the circulating progesterone as a result of the pseudo-ovulation has been mostly related to its critical role in the maintenance of the uterus and embryo development up to the end of gestation since by this time most embryos are being resorbed through a natural selective abortion process [10, 14]. Nevertheless, although progesterone fluctuates during gestation, its levels might be enough to induce extensive side-branching and alveologenesis in mammary glands of pregnant vizcachas.

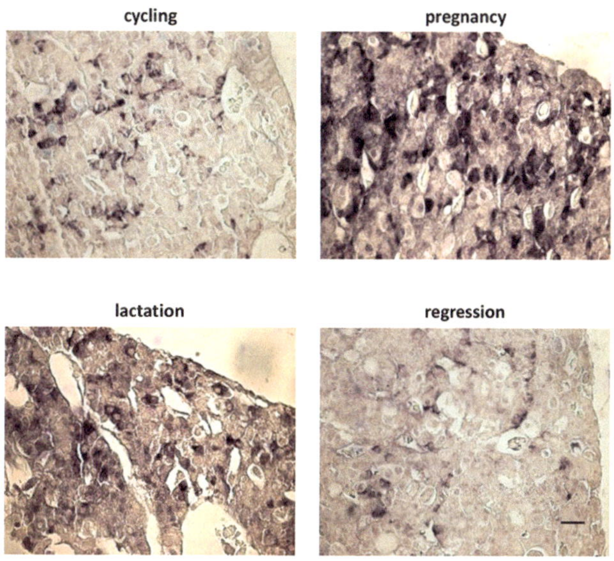


**Figure 4.** PRL hypophyseal content according to the developmental status of the mammary gland of female vizcachas. Representative photomicrographs of adenohypophysis sections of adult vizcachas whose mammary glands are at cycling, pregnancy, lactation and regression status. Immunoreactivity is shown in black. All photomicrographs have the same magnification. Scale bar is depicted in the last photo (bottom right) and represents 25 μm.

Right before parturition, alveolar epithelial cells are enlarged due to a high content of milk fat globules. These alveoli will ultimately become milk-secreting lobules during lactation. As expected along this reproductive stage, the expression of PRLR in the secretory alveolar cells of mammary glands strongly increases in tune with the hypophyseal PRL content of pregnant vizcachas (**Figures 3** and **4**) [15, 26, 27]. On the other hand, even though it has been described that PRL regulates mammary epithelial cell proliferation also via autocrine/paracrine mechanisms [28, 29], we could not detect PRL expression in mammary glands of *L. maximus* neither at protein nor at mRNA level (not shown).

Interestingly, our data shows that, at the peri-pseudo-ovulation interval (approximately between days 90 and 100 of gestation), circulating estradiol peaks and both ERα and ERβ increase their expression in mammary glands (**Figure 3**). ERα localizes in nuclei of both secretory epithelia and stromal cells located immediately beneath of it, supporting the idea of a paracrine role for this transcription factor [17]. Moreover, these data correlate with the accelerated ductal proliferation, branching and alveolar differentiation of mammary glands toward the end of gestation [15]. It has been described that besides its role in pubertal branching, ER$\alpha$ is also essential in alveologenesis during pregnancy and lactation [30]. As for ER$\beta$, it has been reported its requirement for normal lobuloalveolar development during pregnancy rather than for prepubertal growth [31].

Both PRL and luteinizing hormone (LH) are intimately linked to estradiol expression. As result of the hypothalamic-hypophyseal-gonadal axis re-activation in adult pregnant vizcachas, serum LH significantly raises, targets the ovaries and triggers pseudo-ovulation. From there and up to the end of pregnancy, whereas LH gradually decreases, hypophyseal PRL concentration progressively increases up to parturition and remains high during lactation. It has been demonstrated that estrogens target lactotrophs and stimulate PRL gene expression and release, enhance storage capacity and increase cell proliferation [32]. Our preliminary results in adenohypophysis of vizcacha show that hypophyseal ER$\alpha$ is highly expressed at term-gestating females [33]. Last but not the least, at the time of pseudo-ovulation, expression of both hypothalamic PR and gonadotropin-releasing hormone (GnRH) markedly increases. This strongly suggests a role of the hypothalamic-hypophyseal-gonadal axis in the modulation of ovulation during gestation in *L. maximus* [13]. Knowing that ovarian hormones are key players in adult mammary gland growth and development, we could hypothesize that GnRH may play an indirect role in mammary gland remodeling. Moreover, in the near future, we should direct our efforts to elucidate ER$\alpha$ modulation over hypophyseal PRL expression in both pregnant and lactating vizcachas.

### 3.3. Lactation

At this stage, milk-secreting alveoli occupy most of the lobule in the mammary glands of *L. maximus*. As late pregnancy transitioned to lactation, mammary glands consist almost completely of secretory epithelium forming the alveolar structures with lumens full of milk fat globules and milk (**Figure 3**). The magnitude of the dramatic change in the mammary gland architecture is pointed out by the difference in mammary gland weight and size. It has been already described that the fully developed lactating mammary gland in a mouse is seven to 10 times heavier than the mature virgin gland [34].

The secretory epithelial cells of mammary glands during the lactation phase are cuboidal and visibly polarized. The cell nucleus is positioned basally, and the cytoplasm is vacuolated and full of milk droplets. The lumen of alveoli and ducts are full of milk as well. The contraction of myoepithelial cells that surround alveoli helps to empty their content into the interlobular ducts. A very thin connective tissue sheath surrounds each alveolus. We observed the presence of immune cells in the stromal connective tissue and within the milk into the alveoli and ducts. No differences were observed in the morphology between anterior and posterior mammary glands. Anterior and posterior glands are highly branched and full of milk. In fact, we observed that pup suckling occurs indistinctly among the nipples. Lactating females exhibit only one milk patch beneath the skin along the milk line that contains both anterior and posterior nipples [15].

PRL has been well characterized as a terminal differentiation factor of the mammary epithelial cells and for synthesis of milk components during lactation [35]. While mammary glands of *L. maximus* go through a lactation phase, PRLR alveolar expression reaches its highest level which correlates with a high content of hypophyseal PRL (**Figures 3** and **4**).

During lactation, mammary gland expression of PR is much stronger than in any other reproductive state and such expression shifts to the cytoplasm of alveolar cells although some nuclei still show positivity for this receptor (**Figure 3**). This could indicate that the PR antibody used in our experiments recognizes both isoforms of PR (PRA and PRB) which have been described co-expressing in mammary glands of mice at late pregnancy [36].

### 3.4. Regression

Weaning of the litter triggers the process of regression, whereby the mammary gland is remodeled back to its pre-pregnancy state. Mammary gland regression is a period of intensive tissue remodeling. During milk stasis, mammary gland epithelial cells change from a secretory cuboidal to a nonsecretory squamous epithelium. One of the aspects that characterized this stage in *L. maximus* is the detachment of alveolar epithelial cells that shed into the lumen. The structure of the gland displays major changes: alveoli start to collapse, basement membrane becomes fragmented and connective tissue, mostly fibroblast and some adipocytes, start to refill (**Figure 3**). Apoptotic cells, cellular debris and milk components must be cleared for normal regression to proceeds. It is notorious the presence of polymorphonuclear cells in the stroma, infiltrated in the secretory epithelia and in the lumen of the alveoli and ducts of regressing mammary glands of *L. maximus* [15]. Interestingly, it has been described that besides the classical phagocytosis carried out by macrophages, "nonprofessional phagocytes" such as epithelial cells, endothelial cells and fibroblasts also have the capability to participate in the removal of neighboring cells that have undergone apoptosis [34].

These mechanisms that ultimately lead to the regression of the gland are not synchronized in the entirety of the gland of vizcachas. Whereas some lobules display their ductal network disorganized and massive epithelial cell death, other lobules still show alveolar epithelial cells with cytoplasmic fat droplets and alveoli and intralobular ducts with milk remains [15]. This is consistent with the fact that, in natural involution, pups will continue to suckle intermittently as they move to a solid diet. Therefore, in natural involution, mammary gland remodeling proceeds in an unsynchronized fashion with different areas of the gland undergoing involution at different times [34].

The values of circulating ovarian hormones and the expression of their receptors in regressing mammary glands of *L. maximus* notoriously decrease compared to lactation and pregnancy stages. It is almost as if it were a necessary condition to allow mammary gland to go through the remodeling associated with this stage. Strikingly, our preliminary data show that GnRH content at medial basal hypothalamus is higher during the regression stage compared to full term pregnant vizcachas ($1.2 \pm 0.1$ and $0.48 \pm 0.08$ pg/μg total proteins, respectively). This is very interesting considering a recent report published by Rieanrakwong and col. [37] that shows that involution is also dependent on mammary gonadotropin-releasing hormone expression that is suppressed by PRL during lactation.

## 4. Concluding remarks

Although other rodents, such as mice and rats, show an enhanced mammary gland development toward the end of gestation, plains vizcachas also exhibit a pseudo-ovulation event at midterm that causes a sharp rise in circulating progesterone and estradiol which correlates with an augment in the expression of ER$\alpha$, ER$\beta$ and PRLR in mammary glands. These events correlate with the development of a more elaborated and differentiated ductal network and pinpoint a possible relation between the hypothalamic-hypophyseal-gonadal reactivation axis at mid-gestation and the accelerated mammary gland branching and alveolar differentiation of *L. maximus*. Pseudo-ovulation at mid-gestation, which is thought to rescue distal fetuses from selective abortion, influences a precocious development of the mammary gland, preparing females to face the nutritional demand of fully developed newborn in this seasonal-breeding species.

## Acknowledgements

This work was supported by CONICET (PIP N° 0272) and by Fundación Científica Felipe Fiorellino, Universidad Maimónides, Argentina. We are especially grateful to the Ministerio de Asuntos Agrarios, Dirección de Flora y Fauna, Buenos Aires Province for enabling animal capture, to the personnel of ECAS for their invaluable help in trapping and handling the animals, to VMD. Sergio Ferraris and his staff for their essential help on vizcachas handling and anesthetizing, to Mr. Alejandro Schmidt for his technical assistance in image processing and to Ms. María Grisel Clausi Schettini for her excellent technical assistance in tissue processing.

## Author details

Julia Halperin[1,2*], Veronica B. Dorfman[1,2] and Alfredo D. Vitullo[1,2]

*Address all correspondence to: halperin.julia@maimonides.edu

1 Center for Biomedical, Biotechnological, Environmental and Diagnostical Studies (CEBBAD), Maimónides University, Buenos Aires, Argentina

2 CONICET, Buenos Aires, Argentina

## References

[1] Capuco AV, Akers RM. The origin and evolution of lactation. J Biol 2009; 8(4):37.

[2] Parmar H, Cunha GR. Epithelial-stromal interactions in the mouse and human mammary gland in vivo. Endocr Relat Cancer 2004; 11:437–458.

[3] Borowsky AD. Choosing a mouse model: experimental biology in context — the utility and limitations of mouse models of breast cancer. Cold Spring Harb Perspect Biol 2011; 3:a009670.

[4] Keightley MC, Fuller PJ. Anomalies in the endocrine axes of the guinea pig: relevance to human physiology and disease. Endocr Rev 1996; 17:30–44.

[5] Porter KB, Tsibris JC, Nicosia SV, Murphy JM, O'Brien WF, Rao PS, Spellacy WN. Estrogen-lnduced guinea pig model for uterine leiomyomas: do the ovaries protect? Biol Reprod 1995; 52:824–832.

[6] Liehr JG. Hormone-associated cancer: mechanistic similarities between human breast cancer and estrogen-induced kidney carcinogenesis in hamsters. Environ Health Perspect 1997; 105:565–569.

[7] Jackson JE, Branch LC, Villarrreal D. *Lagostomus maximus*. Mamm Species 1996; 543:1–6.

[8] Weir BJ. The reproductive physiology of the plains viscacha, *Lagostomus maximus*. J Reprod Fertil 1971; 25:355–363.

[9] Weir BJ. The reproductive organs of the female plains viscacha, *Lagostomus maximus*. J Reprod Fertil 1971; 25:365–373.

[10] Jensen F, Willis MA, Albamonte MS, Espinosa MB, Vitullo AD. Naturally suppressed apoptosis prevents follicular atresia and oocyte reserve decline in the adult ovary of *Lagostomus maximus* (Rodentia, Caviomorpha). Reproduction 2006; 132:301–308.

[11] Jensen F, Willis MA, Leopardo NP, Espinosa MB, Vitullo AD. The ovary of the gestating South American plains vizcacha (*Lagostomus maximus*): suppressed apoptosis and corpora lutea persistence. Biol Reprod 2008; 79:240–246.

[12] Leopardo NP, Jensen F, Willis MA, Espinosa MB, Vitullo AD. The developing ovary of the South American plains vizcacha, *Lagostomus maximus* (Mammalia, Rodentia): massive proliferation with no sign of apoptosis-mediated germ cell attrition. Reproduction 2011; 141:633–641.

[13] Dorfman VB, Saucedo L, Di Giorgio NP, Inserra PI, Fraunhoffer N, Leopardo NP, Halperín J, Lux-Lantos V, Vitullo AD. Variation in progesterone receptors and GnRH expression in the hypothalamus of the pregnant South American plains vizcacha, *Lagostomus maximus* (Mammalia, Rodentia). Biol Reprod 2013; 89(5):115.

[14] Dorfman VB, Fraunhoffer N, Inserra PI, Loidl CF, Vitullo AD. Histological characterization of gonadotropin-releasing hormone (GnRH) in the hypothalamus of the South American plains vizcacha (*Lagostomus maximus*). J Mol Histol 2011; 42:311–321.

[15] Halperin J, Dorfman VB, Fraunhoffer N, Vitullo AD. Estradiol, progesterone and prolactin modulate mammary gland morphogenesis in adult female plains vizcacha (*Lagostomus maximus*). J Mol Histol 2013; 44(3):299–310.

[16] Bocchinfuso WP, Lindzey JK, Hewitt SC, Clark JA, Myers PH, Cooper R, Korach KS. Induction of mammary gland development in estrogen receptor-alpha knockout mice. Endocrinology 2000; 141:2982–2994.

[17] Mallepell S, Krust A, Chambon P, Brisken C. Paracrine signaling through the epithelial estrogen receptor a is required for proliferation and morphogenesis in the mammary gland. Proc Natl Acad Sci USA 2006; 103:2196–2201.

[18] Cheng G, Weihua Z, Warner M, Gustafsson JA. Estrogen receptors ER alpha and ER beta in proliferation in the rodent mammary gland. Proc Natl Acad Sci USA 2004; 101:3739–3746.

[19] Ben-Jonhatan N, LaPensee CR, LaPensee EW. What can we learn from rodents about prolactin in humans? Endocr Rev 2008; 29(1):1–41.

[20] Ormandy CJ, Horseman ND, Naylor MJ, Harris J, Robertson F, Binart N, Kelly PA. Mammary gland development. In: Horseman ND, editors. Prolactin. Boston, MA: Kluwer; 2001. P. 220–232.

[21] Clevenger CV. Nuclear localization and function of polypeptide ligands and their receptors: a new paradigm for hormone specificity within the mammary gland? Breast Cancer Res 2003; 5: 181–187.

[22] Fiorillo AA, Medler TR, Feeney YB, Wetz SM, Tommerdahl KL, Clevenger CV. The prolactin receptor transactivation domain is associated with steroid hormone receptor expression and malignant progression of breast cancer. Am J Pathol 2013; 182(1):217–33. doi:10.1016/j.ajpath.2012.09.021.

[23] Kavarthapu R, Tsai Morris CH, Dufau ML. Prolactin induces up-regulation of its cognate receptor in breast cancer cells via transcriptional activation of its generic promoter by cross-talk between ER$\alpha$ and STAT5. Oncotarget 2014; 5(19):9079–91.

[24] Macias H, Hinck L. Mammary gland development. Wiley Interdiscip Rev Dev Biol 2012; 1(4):533–557.

[25] Fraunhoffer N. Renovation of ovarian germinal mass in the poliovulatory animal model (Lagostomus maximus) and in the ovarian human endometriosis. [PhD Thesis]. Buenos Aires: Universidad Maimonides; 2015.

[26] Corso MC, Proietto S, Inserra PIF, Charif SE, Dorfman VB, Vitullo AD, Halperin J. Correlations between prolactin, progesterone and VEGF with the level of mammary

http://dx.doi.org/10.5772/66155

gland cellular proliferation along the reproductive cycle of Lagostomus maximus (Rodentia: Chinchillidae). Medicina 2015; 75(II):285.

[27] Proietto S, Corso MC, Schmidt AR, Charif SE, Inserra PIF, Vitullo AD, Dorfman VB, Halperin J. Prolactin negatively correlates with its ovarian receptor at the end of gestation of Lagostomus maximus (Rodentia: chinchillidae). Medicina 2015; 75 (II):228.

[28] Clevenger CV, Chang WP, Pasha TL, Montone KT, Tomaszewski. Expression of prolactin and prolactin receptor in human breast carcinoma. Evidence for an autocrine/paracrine loop. Am J Pathol 1995; 146(3): 695–705.

[29] Naylor MJ, Lockefeer JA, Horseman ND, Ormandy CJ. Prolactin regulates mammary epithelial cell proliferation via autocrine/paracrine mechanism. Endocrine 2003; 20(1–2):111–4.

[30] Feng Y, Manka D, Wagner KU, Khan SA. Estrogen receptor-$\alpha$ expression in the mammary epithelium is required for ductal and alveolar morphogenesis in mice. Proc Natl Acad Sci USA 2007; 104:14718–14723.

[31] Förster C, Mäkela S, Wärri A, Kietz S, Becker D, Hultenby K, Warner M, Gustafsson JA. Involvement of estrogen receptor beta in terminal differentiation of mammary gland epithelium. Proc Natl Acad Sci USA 2002; 99:15578–15583.

[32] Ben Jonathan N, Hnasko R. Dopamine as a prolactin (PRL) inhibitor. Endocr Rev 2001; 22:724–763.

[33] Yankelevich LS, Charif SE, Inserra PIF, Proietto S, Di Giorgio NP, Corso MC, Ochman D, Lux-Lantos V, Halperin J, Vitullo AD, Dorfman VB. Estrogen effect over LH and FSH pituitary expression and its relation with the ovulatory event at mid-gestation of South America plains vizcacha (Lagostomus maximus). Medicina 2015; 75 (II):541.

[34] Atabai K, Sheppard D, Werb Z. Roles of the innate immune system in mammary gland remodeling during involution. J Mammary Gland Biol Neoplasia 2007; 12(1): 37–45.

[35] Ormandy CJ, Naylor M, Harris J, Robertson F, Horseman ND, Lindeman GJ, Visvader J, Kelly PA. Investigation of the transcriptional changes underlying functional defects in the mammary glands of prolactin receptor knockout mice. Recent Prog Horm Res 2003; 58:297–323.

[36] Shyamala G, Yang X, Cardiff RD, Dale E. Impact of progesterone receptor on cell-fate decisions during mammary gland development. Proc Natl Acad Sci USA 2000; 97:3044–3049.

[37] Rieanrakwong D, Laoharatchatathanin T, Terashima R, Yonezawa T, Kurusu S, Hasegawa Y, Kawaminami M. Prolactin suppression of gonadotropin-releasing hormone initiation of mammary gland involution in female rats. Endocrinology 2016; 157(7):2750–8.

Chapter 3

# Role of Melatonin and the Biological Clock in Regulating Lactation in Seasonal Sheep

Edyta Molik and Dorota Zięba-Przybylska

Additional information is available at the end of the chapter

http://dx.doi.org/10.5772/66208

**Abstract**

Impact of light on animal behavior has been known for a long time—from 1925, Rowan [30] showed that lighting conditions influence gonad activity in birds and the related processes are controlled not only by means of intraorganic signals. Studies carried out in subsequent years have established that, also in mammals, the gland reacting to changes in light conditions is the pineal gland, producing a substance called melatonin. Biosynthesis of melatonin in most animals studied to date occurs at a rhythm dependent on the photocycle. The highest concentrations of this hormone—often called "the hormone of darkness"—are recorded at night. Seasonal changes in melatonin secretion conditioned by activity of the biological clock, known also as "biochemical calendar", are the key signals in the annual reproductive cycles of animals exhibiting seasonality of reproduction. Seasonality in sheep refers not only to the reproduction itself but also to lactation. One of the main hormones conditioning initiation and maintenance of lactation, synthesis of milk proteins, fat and immunoglobulins is prolactin (PRL), secreted primarily by lactotrophic cells in the adenohypophysis. Prolactin is also produced locally by the mammary gland—the hormone of this origin is identical to prolactin secreted by the pituitary gland. Until now, it was considered that the level of milk production in mammals is determined by both genetic and environmental factors. However, in recent years, many studies focused on the role of light as a modulator of prolactin levels. In livestock, changes in light-period length play a very important role as this determines their productivity and milk yield. Photoperiod is particularly important in short-day breeder animals (sheep), for which the length of light period is associated with changes in melatonin level. The modulating effect of melatonin on secretion of prolactin may take place via two different mechanisms. One is associated with the circadian rhythm, wherein—directly or through the medium of a factor popularly termed "tuberalin"—melatonin stimulates the release of prolactin. However, this effect is short-lived and is most likely applicable only to prolactin stored in lactotrophic cells of the pituitary. The second mechanism regulating the secretion of melatonin and prolactin is associated with the annual rhythms of secretion—melatonin, due to its lipophilic characteristics, has a direct effect on the secretion of prolactin. Under natural conditions, the maximum concentration of prolactin in the blood of sheep is observed over the long-day period, during which the melatonin level decreases. The lowest prolactin concentration is observed over the short-day period, where melatonin levels are at their highest. Changes in secretion of prolactin during lactation in sheep undoubtedly affect the amount of milk produced.

**Keywords:** seasonality sheep, melatonin, length days, biological clock lactotropic and metabolic hormone

# 1. Introduction

## 1.1. Genetic predispositions and impact of environmental factors on the lactation process in sheep

### *1.1.1. Role of melatonin and the biological clock*

Sheep are short-day breeders, in which the signal for onset of estrus occurs after the summer solstice and is maintained until winter. Such a model of the reproductive cycle in which the young ones are born in the spring provides favorable conditions for rearing lambs as this period coincides with the time of abundance of food, thereby the young have time to put aside fat for the winter. Seasonality of reproductive cycle in sheep is associated with the season and the day length as a recurring reproductive cycle is an endogenous rhythm, encoded genetically. Information on changes in photoperiod reaches the animal's organism through a multineural tract. Much more studies confirm the presence of a molecular mechanism—located in the SCN (suprachiasmatic nucleus) as well as in pars tuberalis (PT)—involved in decoding the melatonin signal.

Both the SCN and PT host over a dozen genes of the circadian clock, such as *Bmal1, Clock, Per1, Per2, Cry1, Cry2A, Rev-erbα* and *CK1ε*, which are mutually coupled [1, 2]. Most likely, the summer and winter rhythm observable in sheep is conditioned by the biological clock genes. Over 24 hours, changes in the melatonin profile affect the rhythmic changes in the expression of these genes as evidenced by varying levels of clock genes mRNA in the PT and SCN. The peak gene expression of *Cry1* (Cryptochrome) gene occurs at twilight, together with the increase in melatonin level, while expression of the gene *Per1* (Period) is stimulated by approaching dawn [3, 4]. In contrast to *Cry1, Cry2* gene is not melatonin-induced [5]. On the other hand, expression of the *Per1* gene is melatonin-dependent as pinealectomy (surgical removal of the pineal gland) blocks the rhythm of the *Per1* gene in the PT, but does not affect the expression of this gene in the SCN.

Studies have reported that repeated multiple injections of melatonin in animals previously subjected to pinealectomy restore the cyclical transcription of *Per1* in PT [6]. The results of studies on biological clock genes in mammalian SCN showed that the BMAL1/CLOCK protein complex encoded by the genes of *Bmal1*, *Clock* induces activation of the *Per1*, *Per2*, *Cry1*, *Cry2*, *Rev-erbα* genes [2]. Studies in sheep artificially subjected to a sudden light stimulus characteristic of a long-day period have shown that the gene expression profile of *Cry1* and *Per1* mirrored that occurring in natural conditions, i.e., expression of *Cry1* increased at night, while that of *Per1* was rising during the day, indicating the presence of a rapid mechanism for regulation of *Cry1* and *Per1* gene expression in response to a melatonin impulse [1].

The neurosensory receptor of circadian rhythm in mammals is the retina of the eye, through which light stimuli are transmitted to the suprachiasmatic nucleus of the hypothalamus (SCN). The pathway the light stimulus travels from the retina to the SCN is known as the retinohypothalamic tract. Secretion of melatonin is a biochemical signal informing the body about changes occurring in the external environment. Melatonin has lipophilic properties and is secreted from the pineal gland by simple diffusion [7]. Because of a well-developed network of blood vessels in the pineal gland, this hormone is released directly into the blood and distributed throughout the entire organism. In animals sensitive to changes in day length, the melatonin profile is a biochemical signal regulating the processes of reproduction and lactation [8]. In sheep, which are short day breeders, seasonal changes in melatonin levels inform the fetus of the environmental conditions.

A large number of $MT_1$ melatonin receptors are present in pars tuberalis (PT) of sheep, while no similar high concentration is noted in the tuberal region of the hypothalamus and in the SCN. This suggests that melatonin may modulate the secretion of hormones secreted only in the pars tuberalis [9, 10]. Increase in melatonin secretion in sheep is stimulated already within 1 or 2 hours after sunset and lasts until the onset of dawn. According to Misztal et al. (1999) [11], the modulating effect of melatonin on prolactin (PRL) secretion could be explained by two different mechanisms. One is linked to the circadian rhythm, which may either have direct impact or act through the factor conventionally known as tuberalin. However, this effect is short-lived and most likely is only applicable to prolactin stored in the lactotropic cells of the pituitary. It is possible that tuberalin activates the expression of the prolactin gene in lactotropic cells [9]. The second mechanism modulating melatonin secretion is related to the annual rhythm of secretion—this means that melatonin, due to its lipophilic properties, has a direct effect on the lactotrophic cells of the pituitary and thus also impacts the secretion of prolactin [11, 12]

### 1.1.2. *Role of prolactin and the growth hormone (GH)*

Changes in melatonin and prolactin profiles in sheep are closely interlinked. Regulation of PRL secretion by melatonin may occur via two different mechanisms. In the case of the circadian rhythm mechanism, melatonin may directly affect PRL secretion—this option applies to the hormone stored in the lactotropic cells; the process may also be mediated by the aforementioned peptide—tuberalin, which activates PRL gene expression in lactotropic cells of the anterior pituitary [9]. In contrast, the process of annual PRL secretion rhythm is directly induced by melatonin that—due to its lipophilic nature—affects the lactotropes [13]. Synthesis of prolactin occurs in the anterior pituitary in the lactotropic cells. The main role of PRL is to initiate and control processes such as mammogenesis, lactogenesis, galactopoesis and involution. Moreover, this hormone plays an important role in biosynthesis of milk proteins (β-casein, α-lactalbumin, lactose). It has been shown that in sheep, the daily and seasonal rhythms of PRL and melatonin are characterized by volatility. Changes in prolactin levels throughout the day are strongly associated with the season. Increases in PRL concentration in the spring are observed at dawn and before dusk. In the summer, the peak level is recorded halfway through the dark cycle, while the autumn PRL profile is characterized by a spike in the first half of the photoperiod and near its end. In short-day breeders, such as sheep, the seasonal lengthening of day-light hours (spring, summer) resulting in a short melatonin signal (4–8 hours a day) does not inhibit the secretion of PRL, while in autumn and winter, a long-lasting melatonin impulse (>10 hours per day) causes a decrease in prolactin concentration [6, 14]. Melatonin modulates PRL secretion also through the intermediary of dopamine. The neurotransmitter stimulates PRL secretion acting through dopamine D1 receptors and inhibits the secretion of this hormone via its effect on dopamine D2 receptors. The presence of seasonal changes in melatonin and prolactin profiles was also confirmed by tests carried out on sheep kept for dairy purposes. It has been shown that key factors affecting milk yield in ewes are changes in the photoperiod. It has been found that milk yield in females entering lactation during the day-light lengthening season is by far (50%) higher than that in animals starting milk synthesis in short-day conditions [15]. The reaction to the shortening of the photoperiod was an increase in melatonin levels, decrease in PRL concentration and lower milk yield. Subjecting sheep to conditions of artificially prolonged day-light cycle (16L:8D) resulted in light-induced inhibition of melatonin synthesis in the pineal gland. Decrease in PRL concentration and lower milk yield were observed simultaneously. Thus, the artificial prolongation of the photoperiod during short-day season is not enough to maintain lactation in seasonal sheep [10, 14, 16, 17]. Previous observations indicate that such processes as mammogenesis, lactogenesis, galactopoesis and involution in sheep require the presence of multiple factors, strongly interdependent. Milk production is based on the impact of a number of factors and day length, as well as changes in PRL profile are only some of many [18]. An important role is also played by the somatotropic system (GH, IGF-1). The growth hormone, similar to PRL, is produced in the anterior pituitary and is involved in synthesis of proteins and fatty acids; it also lowers the concentration of glucose in the blood and is partly responsible for synthesis and secretion of prolactin [19]. Increase in concentration of the GH is stimulated by the "suckling factor"—higher growth hormone levels are observed at the beginning of lactation. Studies carried out on sheep have shown that changes in concentration of both the GH and PRL are dependent on the length of day and are linked to changes in the profile of pineal melatonin. Periodic changes in melatonin levels result in rhythmic inhibition of PRL secretion [20]. It is known that during lactation, under the impact of suckling, the GH and PRL levels in the blood are boosted [19]. Increase in GH secretion during lactation is controlled by GHRH and endogenous opioids. Recently, attention has been drawn to a compound, derivative of dopamine, known as salsolinol it has been shown that concentration of this substance increases in the case of various dysfunctions in the dopaminergic system. Salsolinol stimulates the release of PRL in rodents and ruminants. In lactating sheep, the presence of salsolinol was confirmed in MBH and increase in its concentration in response to suckling was recorded [21]. Salsolinol administered to the third ventricle of the brain during lactation increases prolactin concentration.

Salsolinol antagonist is a compound called 1-MeDIG—this substance inhibits the release of PRL and cancels the stimulatory impact of the "suckling factor" on PRL secretion in rats. A similar effect was observed in sheep in which 1-MeDIQ acts directly on the central nervous system [22]. The compound inhibits the increase in the level of noradrenaline (NA), which acts as a mediator between salsolinol and GH. Interestingly, 1-MeDIQ does not affect the changes in GH concentration induced by stimulation of the mammary gland during suckling. In sheep, both over the period of lamb rearing and beyond, PRL levels decreased after 1-MeDIG was administered. It was proven experimentally that salsolinol has no direct influence on GH profile during lamb rearing. However, in rats subjected to simultaneous administration of both salsolinol and 1-MeDIQ, no statistically significant changes in pituitary hormone levels were observed with the exception of prolactin.

### *1.1.3. Role of metabolic hormones*

The role of metabolic hormones in the process of lactation has garnered a lot of attention. An important role in initiating and maintaining lactation in small ruminants is played by the thyroid hormones, ghrelin and orexin. The production capacity of animals (milk yield, growth and development, coat growth) is largely dependent on proper functioning of the thyroid hormones. These hormones influence also the processes of reproduction in many species of animals, including sheep and goats [23]. In seasonal species, T4 and T3 are obligatory for the annually recurring termination of reproductive activity [24, 25]. Thyroxine in sheep with normally functioning thyroid will shorten the reproductive period and quicken the transition into anoestrus. Thyroxine level peaks in ewes in early pregnancy and decreases just before lambing and after the offspring is born. The level of thyroid hormones in sheep varies throughout lactation. At the start of the process, concentration of these substances is low [26]; however, over time, the thyroxine concentration increases. It has been shown that thyroid hormones, especially tri-iodothyronine, have a suppressive effect on expression of the prolactin gene, which can translate into milk yield as well. An important role in lactation belongs to calcitonin and the parathyroid hormone, responsible for modulation of phosphorus (P) and calcium (Ca) levels. Concentration of these elements in milk has major impact on the chemical composition of the product. The presence of the parathyroid hormone is essential for calcium absorption from the gastrointestinal tract; it also enhances synthesis of active D3 (1,25-di-hydroxycalciferol) that stimulates the process of calcium binding by proteins. It has been experimentally demonstrated that thyroid hormone secretion is correlated with day length in sheep. In vitro studies in thyroid gland explants showed higher levels of thyroxine under short-day conditions and lower in the season of elongating photoperiod (spring), while T3 reached higher levels in the summer and lower when the photoperiod was shortening. In addition, there was an increase in T3 concentration induced by exogenous melatonin [25]. Productivity of the animals depends not only on the level of nutrition, environmental factors and their genetic potential. Thyroid hormones are an important link in the key stages of life (reproduction and lactation) of all living organisms [27]. Orexin A is of particular importance in the reproductive process of animals sensitive to changes in day length. The process of initiating and maintaining lactation in sheep requires the presence of many hormones. Defining the role of orexin in regulating their secretion, especially that of prolactin, may allow to better understand the process of maintaining lactation in sheep, in particular over short-day period.

## 2. Influence of day length and melatonin on milking yield

Changes in day length and the related secretion of melatonin and prolactin are of particular significance in sheep as they determine reproductive processes, the last stage of which is lactation. The possibility of artificial extension of the milking period in late-lambing ewes by application of prolonged day length, 16 hours of light—8 hours of darkness (16L:8D), was introduced additionally (Group III). Measurements of plasma levels of prolactin and melatonin were used as parameters of season-dependent hormonal regulation of milk production in this seasonally breeding species [28]. Lambs remained with their mothers up to 56th day of life. Then lambs were separated from their mothers, which were allocated to the milking. During milking period, ewes were milked twice a day using Alfa-Laval machine. Individual milk yield checks were carried out every 10 days. From the 20th day of lactation to the end of this process, the blood samples were carried out from each sheep every 30 days to determine concentration of melatonin and prolactin. Blood sampling started after sunset and continued for 6 consecutive hours with a frequency of every 60 minutes. Blood after collection were centrifuged and the resulting plasma was stored at temperature −20°C until analysis. Hormones have been determined by radioimmunologically (RIA) method. During lambs' rearing period, sheep produced similar quantities of milk, since ewe Group 1 produced 48.2 ± 12.9 liters whereas Group II produced 42.4 ± 16.4 liters. Higher productivity was observed in Group III 60.5 ± 16.6 liters, which was kept in artificial light conditions. The observed differences were statistically insignificant. Distinct differences in milk yield were

| Groups of sheep | Milk yield of the first 28 days of lactation (l) | | Total length of lactation (days) | | Days of milking (days) | | Milk production during milking (l) | |
|---|---|---|---|---|---|---|---|---|
| | $\bar{x}$ | SEM | $\bar{x}$ | SEM | $\bar{x}$ | SEM | $\bar{x}$ | SEM |
| Group I Sheep lambed in January | 48.2 | 2.3 | 177 | 8.6 | 102 | 4.8 | 33.0 | 3.6 |
| Group II Sheep lambed in June | 42.4 | 3.1 | 147 | 3.5 | 77 | 4.0 | 16.8 | 1.4 |
| Group III Sheep lambed in June (16L:8D) | 60.5 | 3.2 | 160 | 4.1 | 90 | 3.9 | 21.2 | 1.7 |

**Table 1.** Parameters characterizing lactation duration and efficiency of Polish Longwool sheep lambing in January (Group I), in June, kept under natural lighting conditions (Group II) and in June, kept under the long-artificial photoperiod (16L:8D, Group III). See text for statistical comparisons.

observed between the groups of sheep milk in the period of use. The highest milk yield of 33.0 ± 11.2 liters was found in Group I, while Group II produced only 16.8 ± 4.4 liters, the obtained differences were statistically significant ($P \leq 0.01$). Mothers who remained in artificial light conditions (Group 3) produced 21.2 ± 5.5 liters of milk (**Table 1**). The results of the total lactation length and days of milking show conclusively that the lactation period in Group I was significantly longer than that in Groups II and III ($P < 0.05$, **Table 1**). Analysis of the course of lactation with regard to the mean amount of milk obtained in particular months of milk use revealed that the milk yield of Group I in the first month of milking (0.43 ± 0.09 liters/day) was similar to the milk yield of Group III (0.42 ± 0.07 liters/day), with only 0.18 ± 0.08 liters/day in Group II ($P < 0.01$, **Figure 1**, **Tables 2–4**).

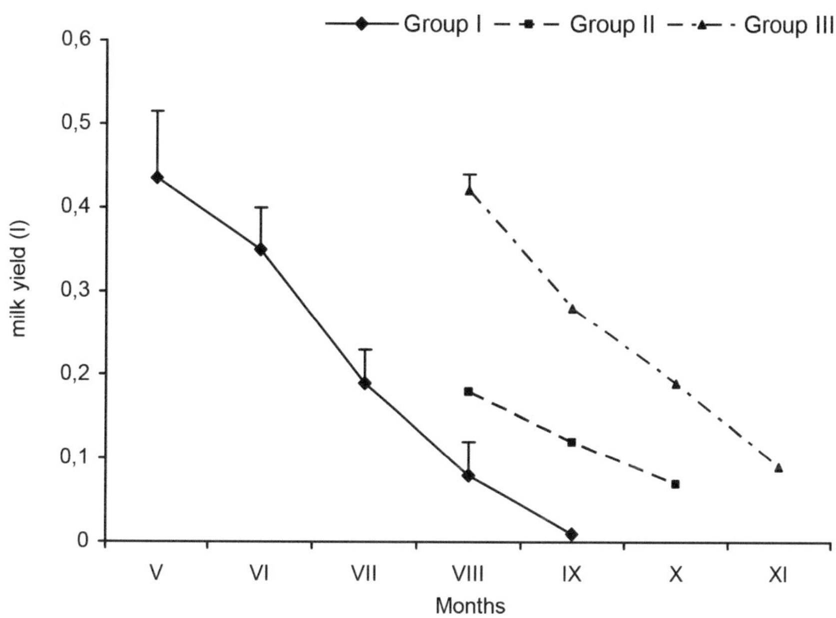


**Figure 1.** Mean monthly milk yield of Polish Longwool sheep lambed in January (Group I), in June, kept under natural lighting conditions (Group II) and in June, kept under the long-artificial photoperiod (16L:8D, Group III), during the milking period. See text for statistical comparisons.

| Months | | II | III | IV | V | VI | VII | VIII | IX |
|---|---|---|---|---|---|---|---|---|---|
| MLT (pg/ml) | $\bar{x}$ | 168.5 | 85.5 | 132.8 | 133.5 | 77. | 73.3[c] | 124.7 | 91.3 |
| | SD | 137.3 | 45.3 | 112.6 | 113.0 | 38.9 | 50.1 | 100.6 | 42.2 |
| PRL (ng/ml) | $\bar{x}$ | 128.6 | 102.8 | 156.5 | 312.6[f] | 185.7 | 247.0[j] | 151.6 | 43.9 |
| | SD | 46.3 | 33.8 | 18.1 | 45.2 | 54.7 | 60.9 | 43.9 | 33.1 |
| Milk (l) | $\bar{x}$ | – | – | – | 0.43 | 0.35 | 0.19 | 0.08 | 0.01 |
| | SD | | | | 0.08 | 0.05 | 0.04 | 0.04 | – |

**Table 2.** Mean (±SEM) and SD of plasma melatonin and prolactin and milk concentrations in sheep lambed in January (Group I). See text for statistical comparisons.

| Months | | VII | VIII | IX | X | XI |
|---|---|---|---|---|---|---|
| MLT (pg/ml) | $\bar{x}$ | 61.1 | 87.8 | 82.3 | 77.5 | 93.2[a] |
| | SD | 28.5 | 45.5 | 45.0 | 46.1 | 57.4 |
| PRL (ng/ml) | $\bar{x}$ | 234.0 | 124.6 | 60.5 | 30.8 | 16.8 |
| | SD | 39.5 | 48.8 | 31.1 | 17.7 | 10.4 |
| Milk (l) | $\bar{x}$ | – | 0.18 | 0.12 | 0.07 | 0.04 |
| | SD | | 0.05 | 0.03 | 0.01 | 0.01 |

**Table 3.** Mean (±SEM) and SD of plasma melatonin and prolactin and milk concentrations in sheep lambed in June kept under natural lighting conditions (Group II). See text for statistical comparisons.

| Months | | VII | VIII | IX | X | XI |
|---|---|---|---|---|---|---|
| MLT (pg/ml) | $\bar{x}$ | 41.0[a] | 60.4 | 17.6 | 4.4 | 17.0 |
| | SD | 19.7 | 39.8 | 27.6 | 4.1 | 15.5 |
| PRL (ng/ml) | $\bar{x}$ | 278.8 | 132.7 | 147.9 | 84.3 | 38.3 |
| | SD | 55.3 | 57.4 | 82.4 | 42.5 | 25.2 |
| Milk (l) | $\bar{x}$ | – | 0.42 | 0.28 | 0.19 | 0.09 |
| | SD | | 0.07 | 0.05 | 0.03 | 0.02 |

**Table 4.** Mean (±SEM) and SD of plasma melatonin and prolactin and milk concentrations in sheep lambed in June and kept under the long-artificial photoperiod (Group III, 16L:8D, bottom). See text for statistical comparisons.

## 2.1. Secretion melatonin and prolactin in day length

The highest melatonin level in Group I determined in February was 168.5 ± 137.3 pg/ml, while prolactin level at this time was 128.6 ± 46.3 ng/ml. The highest prolactin concentration determined in May was 312.6 ± 45.2 ng/ml, further growth of prolactin level in July was 247.0 ± 60.9 ng/ml and the resulting differences were statistically significant ($P \leq 0.05$) compared to the level of the hormone in other administrations. With the lengthening of lactation and changes in day length light from August to September significantly decreased prolactin level from 151.6 ± 43.9 ng/ml to 43.9 ± 33.1 ng/ml and increased levels of melatonin. In August, the concentration of melatonin was the highest and amounted to 124.7 ± 100.6 pg/ml, while sheep milk production has decreased to a level of 0.08 ± 0.02 liters per day. In the last month of lactation, melatonin level was 91.3 ± 42.2 pg/ml and sheep milk production at this time was only 0.01 ± 0.02 liters/day (**Figure 4, Table 4**). The increase in melatonin levels from July to September of 33.4 pg/ml was accompanied by a decrease in prolactin level of 107.7 ng/ml. At that time, there was a decrease in the secretion of milk by an average of 0.11 liters/day (**Table 2**).

## 2.2. Secretion melatonin and prolactin in short days

In the case of Group II, the highest level (234.0 ± 39.5 ng/ml) of prolactin was also determined in July and the melatonin concentration in this period was the lowest (61.1 ± 28.5 ng/ml). With the shortening of the light day, prolactin secretion was decreasing and the level of this hormone was lower by 25.8% in September compared to the level observed in July. A clear decrease in prolactin level observed during the last 2 months of lactation, i.e, October and November was 30.8 ± 17.7 ng/ml and 16.8 ± 10.3 ng/ml, respectively. In July, as previously indicated, the lowest concentration of melatonin was 61.1 ± 28.5 pg/ml, differing significantly ($P \leq 0.05$) to the identified levels of this hormone in November (93.2 ± 57.4 ng/ml). Changes in the concentration of melatonin and prolactin during the shorter photoperiod influenced the parameters of sheep milk production, causing a drop in milk yield by 22.2% between August and November (**Table 3**). The results obtained in Group III showed that the highest level of prolactin found in July was 278.8 ± 55.3 ng/ml and much lower in September, 147.9 ± 82.4 ng/ml; however, it was higher than the level of prolactin identified in August, 132.7 ± 57.4 ng/ml. The resulting differences in the levels of prolactin in the month of July, August and September were statistically significant ($P \leq 0.05$). Despite ensuring that this group of animals underwent 16-hour lighting intensity of 200 lux, the concentration of prolactin from September to November reduced (**Table 4**).

The sheep of all groups produced similar amounts of milk during the first 28 days of lactation as estimated based on the weight gains of the lambs. The study results showed that the shift in lambing date—from winter to summer—had a negative effect on milk production parameters in ewes. Sheep that gave birth in January and were used for dairy purposes over the long-day period produced 50% more milk than ewes that gave birth in June and were then milked as the day length was gradually decreasing. Lactation in sheep milked in the summer-time was significantly longer than that in sheep milked when the photoperiod duration was shortening. Day length had no effect on milk yield in the period of rearing lambs (i.e., 28 days). Monitoring of hormone levels (prolactin and melatonin) in sheep during lactation allowed to conclude that secretion of melatonin in the fall months increased, while prolactin secretion was decreased over the same period. The increase in melatonin level during the shortening of the day in the Polish Longwool sheep reduces prolactin secretion and inhibition of the synthesis of milk. Introduction of artificial light conditions during the shortening of the photoperiod is not enough to maintain secretion of prolactin in ewes at a level that allows to maintain lactation in the autumn.

Length of illuminating day, but especially profile of melatonin has a particular meaning in sheep, because decidate of trial procreative with last stage physiology that last stage physiology reproduction is lactation. [18]. As the many physiological processes also the reproductive cycle is genetically encoded in the sheep. The course of this cycle is reflected by the seasonal changes in the secretory activity of the hypothalamopituitary gonadotropic GnRH-LH system. Sustaining of the proper duration of this cycle requires, however, constant and periodically repeated factors which enable the synchronization of the physiological processes with a suitable season of the year. Thus, the day length plays the most important role in this aspect. The information about the day length reaches the organism as the biochemical signal generated by the pineal gland via the nocturnal secretion of melatonin. The seasonal changes in the duration of melatonin secretion are of great importance in the modulation of sexual activity and lactation in the sheep with the inherent traits of seasonality. The dependence of milk yield and the duration of lactation on melatonin and prolactin secretion are also demonstrated in seasonal and aseasonal breeds of sheep lambed during the different seasons of the year. The putative mechanisms of melatonin action on luteinizing hormone and prolactin secretion are also demonstrated with reference to the melatonin-binding sites in the sheep central nervous system and pituitary gland [29].

### 2.3. Influence of day length and melatonin in prolactin secretion and growth hormone in suckling sheep

The effects of melatonin on the secretion of prolactin (PRL) and growth hormone (GH) were studied in ewes' nursing lambs (Polish Longwool, n = 20) under different photoperiods (March and November). The animals were divided into four groups: (a) (LDC—long-day control group, n = 5), (b) melatonin-treated (LDM—long-day group, n = 5), (c) (SDC—short-day control group, n = 5) and (d) (SDM—short-day melatonin, n = 5). Blood samples were collected from ewes 5 days after lambing. Four blood collections were performed at 10-day intervals, over a 40-day time period. Sampling started at sunset and continued for 6 hours at 20-minute intervals. Melatonin implants (exogenous melatonin) were inserted in ewes of the LDM and SDM groups after first blood collection. The plasma concentrations of PRL and GH were assayed using RIA. In ewes from the LDC group, the mean plasma PRL concentration increased gradually, reaching a significantly ($P < 0.001$) higher level, after 3 weeks. In contrast, in the LDM group, PRL concentration decreased significantly ($P < 0.001$) following 10 days, compared to that in ewes from the LDC group. The mean plasma GH concentration was significantly ($P < 0.001$) higher in the LDC group than that in the LDM group, the for the entire experimental period during the experimental period. In the SDC and SDM groups, plasma PRL concentrations did not decrease significantly ($P < 0.001$) 2 weeks after the onset of the experiment and did not differ significantly between these groups. The mean plasma GH concentration increased significantly ($P < 0.001$) in the SDM group compared with the SDC group only after the third week. The mean plasma GH concentration in the SDM group and the SDC group reached a similar level by the end of the trial. It would appear that melatonin may effectively inhibit PRL secretion in nursing ewes during long photoperiod and stimulate GH release during short photoperiod. The inhibition of PRL secretion in nursing ewes during increasing photoperiod (long days) occurs, despite the strong stimulation of suckling. At the onset of the experiment, the mean plasma PRL concentrations in the LDC and LDM groups were similar (193.2 ± 10.9 and 192.3 ± 8.7 ng/ml, respectively) (**Figure 2**). During the subsequent collection (second), the PRL concentration in the LDC group was 166.2 ± 8.0 ng/ml; however, in the LDM group a significant decrease in the plasma PRL concentration was recorded as 56.5 ± 4.2 ng/ml ($P < 0.001$). During this time, the mean concentration of PRL in LDC ewes was significantly higher ($P < 0.001$) than the LDM ewes (**Figure 2**). The mean concentration continued to increase as day length (photoperiod) increased.

During the decreasing day length period, the PRL secretion profile was similar in SDC and SDM groups (**Figure 2**). In both groups, there was a significant ($P < 0.05$) decrease in plasma of PRL concentration in during the third week of lactation (14.7 ± 1.4 and 12.6 ± 1.0 ng/ml) compared with the initial concentration (60.6 ± 7.3 and 53.4 ± 6.0 ng/ml, respectively). No differences in PRL plasma concentrations were recorded between the SDC and SDM groups.

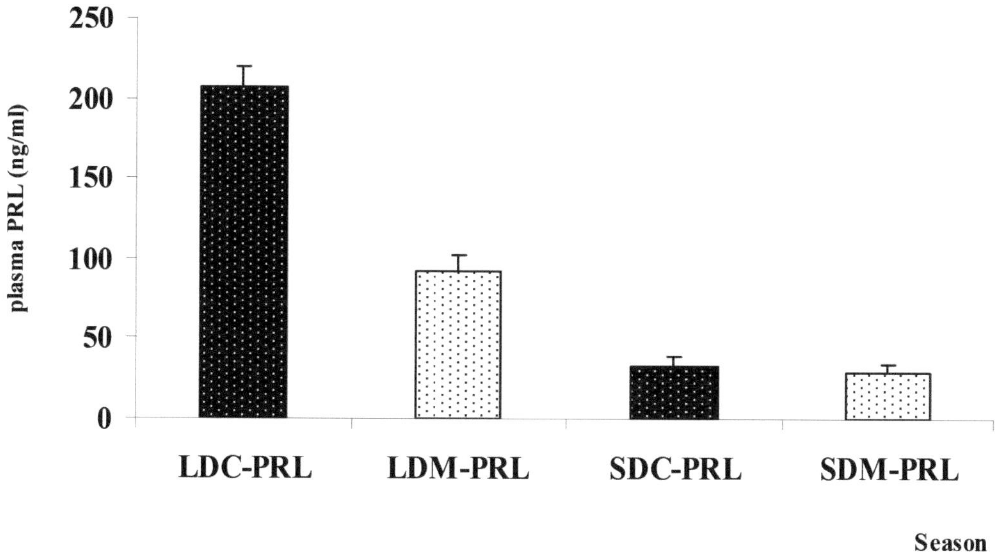


**Figure 2.** Mean plasma PRL concentrations in nursing-sheep lambing (LDC—a long-day control and LDM—a long-day melatonin-treated group, SDC—a short-day control and SDM—a short-day melatonin-treated group). See text for statistical comparisons.

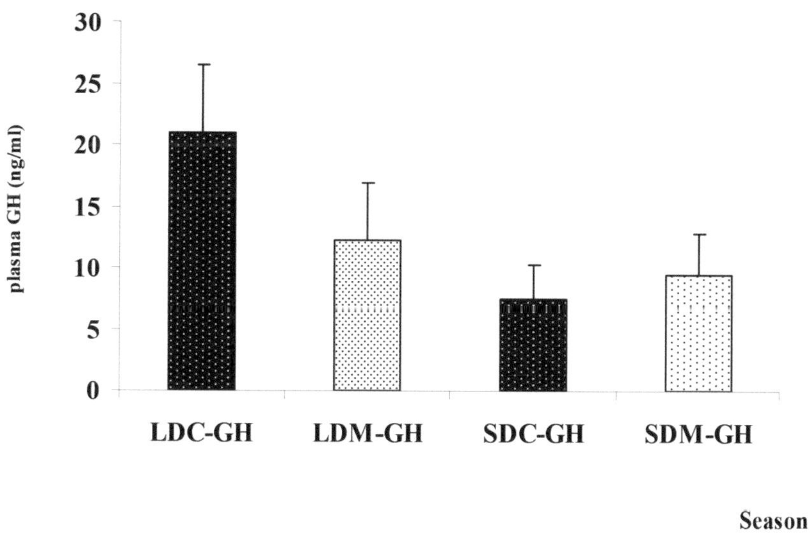


**Figure 3.** Mean plasma PRL concentrations in nursing-sheep lambing (LDC—a long-day control and LDM—a long-day melatonin-treated group, SDC—a short-day control and SDM—a short-day melatonin-treated group). See text for statistical comparisons.

The mean plasma GH concentration was significantly ($P < 0.001$) higher in the LDC group than the LDM group for the entire trial period. It was also observed that a gradual decrease in GH concentration took place in both groups (**Figure 3**). The mean GH concentrations in

SDC and SDM ewes were at a similar level up to the second time of blood collection day (**Figure 3**). The only significant ($P \leq 0.001$) rise in plasma GH secretion was recorded in the third week of lactation in the SDM group (12.23 ± 5.36 ng/ml) compared with the SDC group (6.58 ± 2.36 ng/ml).

In conclusion, the long-term treatment with exogenous melatonin of early-lactating sheep reduced the PRL secretion during the increasing photoperiod, despite strong stimulation by suckling. Moreover, in nursing ewes, melatonin stimulated GH secretion during the short photoperiod. It can therefore be assumed that melatonin may be indirectly affected by the level of milk production in sheep, especially following the nursing period.

## 3. Influence of metabolic hormones on prolactin secretion in lactation sheep

### 3.1. Role of orexin

Studies on the role of orexin A in the control of prolactin (PRL) and growth hormone (GH) secretion in rodents have produced inconsistent results. Orexin A may play a special role in animals' sensitivity such as sheep to the day length changes. The aim of the study was to determine the role of orexin A in the control of prolactin secretion and growth hormone in sheep during different photoperiods. In vitro studies were carried out on 10 Polish Long-wool ewes on 30 days of lactation during long photoperiod (May, LD, n = 5) and short photoperiod (December, SD, n = 5). After rearing lambs to 30 days of age, ewes were decapitated and the pituitaries were dissected and then cut along the longitudinal fissure into two halves, so that each half contained the glandular and nervous parts. Pituitary glands were collected and divided along the longitudinal fissure into two halves. Glands were incubated for 3 hours at 37°C in Parker medium with addition of orexin A—experimental group or in medium alone—control group. During the following 3-hour incubation, medium was exchanged every 15 minutes and a sample of 1 ml was collected and immediately frozen at −80°C until assay. Prolactin concentrations in the medium were determined radioimmunologically (RIA).

In the long-day conditions (May), the pituitary explants of lactating sheep exhibited the strongest secretory activity during the first hour of incubation—significantly higher in orexin-treated group (O1) than the control group (K1), ($P < 0.01$). During the second hour of the incubation, PRL concentration decreased and reached the similar values in both groups. During the third hour, PRL concentration in O1 group was again significantly higher than that noted in K1 group ($P < 0.01$). In the short-day conditions (December), PRL concentration was significantly higher in orexin-treated group O2 during the first hour of incubation than the value observed in the control group—K2 ($P < 0.01$). The inverse relationship in prolactin release was observed during the second hour of incubation ($P < 0.01$), however, during the third hour, PRL concentration was again significantly higher in O2 group than the concentration noted in K2 group ($P < 0.05$). Collective analysis of the data showed that PRL concentrations were higher in experimental groups (O1 and O2) than the concentrations noted in control groups (K1 and K2) under both the long (May) and short (December) photoperiods (**Figure 4**).

http://dx.doi.org/10.5772/66208

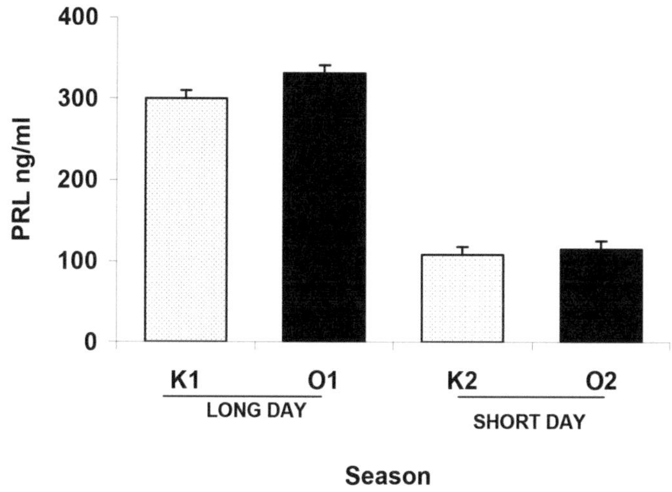


**Figure 4.** Mean concentrations of prolactin in control and orexin A-treated pituitary explant cultures during long-day (LD) and short-day (SD) photoperiods. See text for statistical comparisons.

GH release from the pituitary explants during the long-day conditions was maintained on significantly higher level in orexin-treated group O1 than control K1 group ($P < 0.05$), throughout the whole period of the incubation. In contrast, during the short-day period, GH release from the explants was significantly less in orexin-treated group O2 than that in the control K2 group ($P < 0.05$). The suppressive effect of orexin was observed during 2 hours. Collective analysis of the data showed that GH concentrations were higher under long-day conditions than under short-day conditions (**Figure 5**).

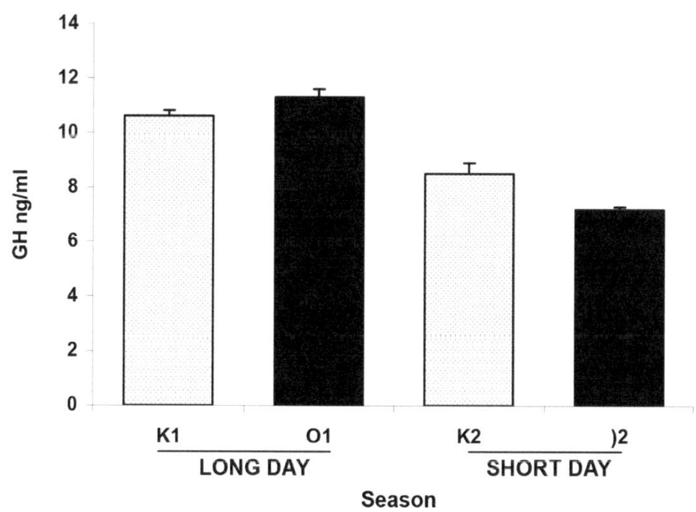


**Figure 5.** Mean concentrations of growth hormone in control and orexin A-treated pituitary explant cultures during long-day (LD) and short-day (SD) photoperiods. See text for statistical comparisons.

The results of experiments performed on lactating sheep, i.e., animals with strong seasonality characteristics, are difficult to compare with others, however, in ewes as in rodents, orexin A is able to stimulate PRL secretion due to its direct effect on the lactotropic cells. The slight response of ovine pituitary glands to orexin during short days is probably due to the insensitivity of lactotropic cells to the orexin signal. The initiation and maintenance of lactation in sheep require the presence of many hormones, where PRL and GH seem to be the most important. Studies on lactating sheep showed that the ewes starting lactation during the period of increasing day length produced 50% more milk compared with sheep milked during the decreasing day length [15]. When June-lambed ewes were kept under artificial conditions of the long day (16L:8D), PRL level decreased as the natural length of a day became shorter. The fact that pituitary cells become refractory to over-repeated summer signal of the darkness hormone (melatonin) makes it impossible to lengthen lactation in sheep in the autumn-winter period [15]. Determining the role of orexins, especially orexin A, in regulating prolactin secretion may help to clarify the process of lactation maintenance in sheep, especially during the decreasing photoperiod. In conclusion, our results obtained on the pituitary explants demonstrated that the pituitary tissue of lactating sheep was sensitive to photoperiod and orexin A. We conclude that the secretion of PRL and GH from the ovine pituitary gland is negatively responsive to orexin A during SD, whereas orexin may stimulate PRL and GH secretion during LD. Further studies investigating orexin—PRL and GH interactions are needed.

### 3.2. Role of TRH

Recently, it was observed that TRH has a role to play in the initiation and maintenance of lactation in small ruminants. The aim of the performed study was to determine the impact of the TRH factor on secretion of prolactin in lactating sheep. In vitro studies were carried out on 10 animals. The pituitary gland of each sheep was collected at day 40 of lactation. In vitro incubations were performed on 12 microwell plates in Parker medium for 1 hour at 37°C. One half of the gland was incubated in pure Parker medium (control group), while the second (test group) half was incubated in Parker medium conditioned with exogenous TRH (TRH concentration—36 ug/100 ml medium). The medium was administered every 15 minutes and collected from the wells; in each case, 1 ml of medium was administered. The first 15 minutes served as blank and both halves of the pituitary remained in the same medium; the aim was to stabilize the secretory function of lactotropic cells. Prolactin measurements were made using RIA method. The tests carried out have demonstrated a stimulating impact of the TRH factor on secretion of prolactin. In the first 15 minutes of incubation, PRL concentration in the control group was 81.83 ± 11.4 pg/ml and was significantly ($P \leq 0.05$) lower than the concentration (87.48 ± 11.6 pg/ml) observed in the test group. After 30 minutes of incubation, the control group showed significantly ($P \leq 0.05$) lower prolactin level (74.04 ± 10.03 pg/ml) than the group with TRH-enriched medium (79.9 ± 10.6 pg/ml). After 45 minutes of incubation, the concentration of PRL in the control group was 59.66 ± 9.4 mg/ml and it was significantly ($P \leq 0.01$) lower than that in the experimental group (10.2 ± 65.47 mg/ml) (**Figure 6**).

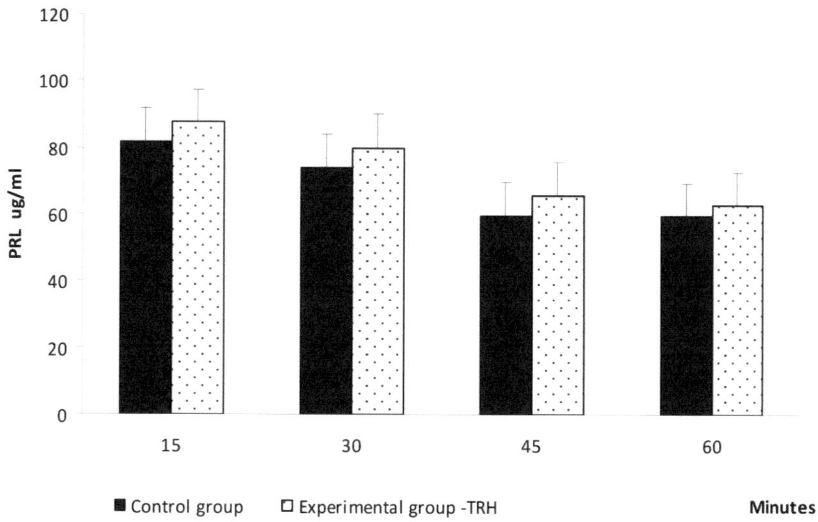


**Figure 6.** Mean concentrations of prolactin in control and TRH-treated pituitary explant cultures during long-day. See text for statistical comparisons.

### 3.3. Role of ghrelin

The process of initiation and maintenance of lactation in sheep requires the presence of a number of hormones. The aim of this study was to determine the role of ghrelin in the regulation of prolactin secretion in lactating sheep, based on the culture of in vitro pituitary. The study was conducted in May—long-day period. Pituitary was collected from 10 sheep on day 30 of lactation and divided along the longitudinal grooves so that each contains half of the glandular part and nerves. Incubations were carried out in vitro pituitary in 12 well plates for 1 hour at 37°C. The control group was incubated in a clean Parker medium and experimented in medium supplemented with exogenous ghrelin. The concentration of prolactin in the medium was determined by RIA method. The study showed stimulatory effect of ghrelin on the secretion of prolactin. The tests demonstrated a modulating effect of ghrelin on secretion of prolactin. Significant ($P \leq 0.05$) increase in prolactin secretion after 30 minutes of incubation in the test group (89.6 ± 18.1 mg/ml) compared with the control group (73.6 ± 17.4 mg/ml) was noted. After 45 minutes of incubation, the concentration (69 ± 15.2 mg/ml) of prolactin in the test group was significantly ($P \leq 0.05$) lower than the concentration (77.2 ± 17.6 mg ml) in the control group. After 60 minutes, prolactin level was significantly lower at $P \leq 0.05$ in the test group (46.3 ± 8.4 mg/ml) than that in the control group (51.8 ± 9.6 mg/ml) (**Figure 7**). The results of studies conducted have demonstrated a modulating impact of ghrelin on secretion of prolactin. While increase in prolactin secretion during the incubation period was observed, reduction in prolactin secretion has been recorded in the test group. Administration of exogenous ghrelin during the period of physiologically high prolactin concentration in lactating sheep has not given a clear answer as to whether ghrelin stimulates the secretion of prolactin. The results suggest, therefore, that ghrelin does not directly affect the secretion of prolactin from the pituitary. The hitherto obtained test results showed that the effects of ghrelin may be dependent on the species of animals. In the case of sheep, seasonal breeders, the mechanism of ghrelin activity is complicated. As revealed by the studies in lactating sheep, administration of exogenous ghrelin modulates the secretion of prolactin.

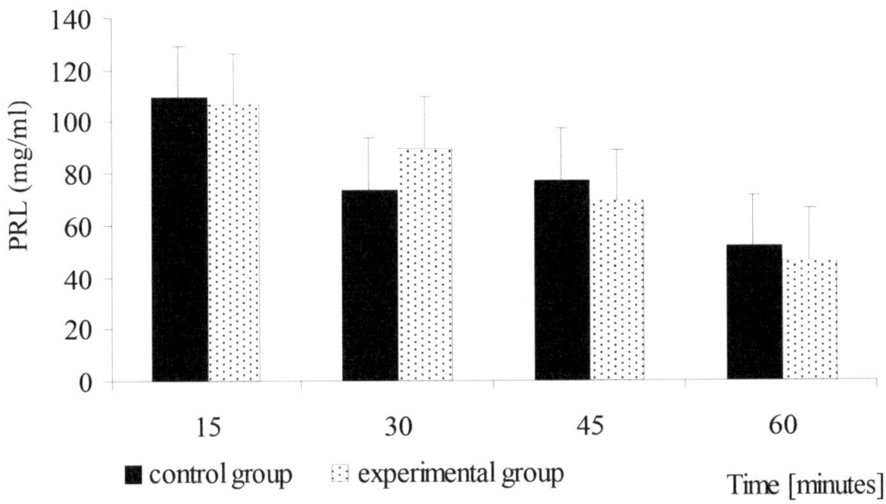


**Figure 7.** Mean concentrations of prolactin in control and ghrelin-treated pituitary explant cultures during long day. See text for statistical comparisons.

## 4. Summary

In seasonal animals, the process of triggering and maintaining lactation requires numerous hormones. The interaction of growth factors and other hormones is necessary in processes such as mammogenesis, lactogenesis and galactopoiesis. Due to the proper synchronization of pregnancy and changes in the area of the mammary gland, the gland is ready for the production of milk at the moment the offspring is born. Mammogenesis is a phenomenon that requires the participation of a number of hormones, including prolactin (PRL), growth hormone (GH), estrogens, progesterone, oxytocin, placental lactogen (PL) and insulin-like growth factor (somatomedin, e.g., IGF1). The coparticipation of IGF and GH is necessary in coordinating the differentiation and proliferation of epithelial cells. The manner in which the growth factors stimulate or inhibit the growth of cells or their influence on the cell cycle is not fully understood. The role of IGF in particular stages of functioning of the mammary gland (mammogenesis, lactogenesis, galactopoiesis and desiccation), particularly in the case of ruminants, is highly complicated. Recently, attention has been given to the metabolic hormones, particularly the role of leptin, orexin and ghrelin in mammogenesis, lactogenesis and galactopoiesis, respectively. Due to the recently increased interest in sheep's milk products, an understanding of the endocrine mechanisms facilitating the maintenance of lactation during autumn and winter may contribute to the improved profitability and usefulness of sheep's milk.

## Author details

Edyta Molik* and Dorota Zięba-Przybylska

*Address all correspondence to: rzmolik@cyf-kr.edu.pl

Department of Animal Biotechnology, Agricultural University in Krakow, Kraków, Poland

## References

[1] Hazlerigg DG andersson H, Johnston JD, Lincoln G: Molecular characterization of the long-day response in the Soay sheep, a seasonal mammal. Current Biology. 2004; 14, 334–339.

[2] Lincoln G, Messager S andersson H, Hazlerigg D: Proceedings of the National Academy of Sciences. Temporal expression of seven clock genes in the suprachiasmatic nucleus and the pars tuberalis of the sheep: evidence for an internal coincidence timer. 2002; 99(21), 13890–13895.

[3] Lincoln GA: Decoding the nightly melatonin signal through circadian clockwork. Molecular and Cellular Endocrinology. 2006; 252, 69–73.

[4] Lincoln GA: Melatonin entrainment of circannual rhythms. Chronobiology International. 2006; 23, 301–306.

[5] Dardente H: Does a melatonin-dependent circadian oscillator in the pars tuberalis drive prolactin seasonal rhythmicity. Journal of Neuroendocrinology. 2007; 19, 657–666.

[6] Lincoln GA, Clarke IJ: Photoperiodically-induced cycles in the secretion of prolactin in hypothalamo-pituitary disconnected rams: evidence for translation of the melatonin signal in the pituitary gland. Journal of Neuroendocrinology. 1994; 6, 251–260.

[7] Cardinali DP, Pévet P: Basic aspects of melatonin action cardinali. Sleep Medicine Reviews. 1998; 2(3), 175–190.

[8] Houghton DC, Young IR, Mcmillen IC: Photoperiodic history and hypothalamic control of prolactin secretion before birth. Endocrinology. 1997; 138, 1506–1511.

[9] Johnston JD: Photoperiodic regulation of prolactin secretion: changes in intra-pituitary signalling and lactotroph heterogeneity. Journal of Endocrinology. 2004; 180, 351–356.

[10] Molik E, Pasternak M, Błasiak M, Misztal T, Romanowicz K, Zieba D: The effect of the diversified signal of melatonin on milk yields in seasonally breeding sheep. Archive fur Tierzucht. 2013; 56(93), 924–929.

[11] Misztal T, Romanowicz K, Barcikowski B: Melatonin modulation of the daily prolactin secretion in intact and ovariectomized ewes. Neuroendocrinology. 1999; 69, 105–112.

[12] Mikolayunas CM, Thomas DL, Dahl GE, Gressley TF, Berger YM: Effect of prepartum photoperiod on milk production and prolactin concentration of dairy ewes. Journal of Dairy Science. 2008; 91(1), 85–90.

[13] Morrissey AD, Cameron AW, Tilbrook AJ: Artificial lighting during winter increases milk yield in dairy ewes. Journal of Dairy Science. 2008; 91(11), 4238–4243. doi: 10.3168/jds.2007-0918.

[14] Misztal T, Romanowicz K, Barcikowski B: Short-term modulation of prolactin secretion by melatonin in anestrous ewes following dopamine - and opiate receptor blockade. Experimental, Clinical, Endocrinology & Diabetes. 2001; 109, 174–180.

[15] Molik E, Misztal T, Romanowicz K, Wierzchoś E: Dependence of the lactation duration and efficiency on the season of lambing in relation to the prolactin and melatonin secretion in ewes. Livestock Science. 2007; 107, 220–226.

[16] Molik E, Misztal T, Romanowicz K, Wierzchoś E: The Influence of length day on melatonin secretion during lactation in aseasonal sheep. Archive fur Tierzucht. 2006; 49, 359–364.

[17] Molik E, Misztal T, Romanowicz K, Zięba D, Wierzchoś E: Changes in growth hormone and prolactin secretion in ewes used for milk under different photoperiodic conditions. Bulletin of the Veterinary Institute in Puławy. 2009; 53, 389–393.

[18] Andrade BR, Salama AA, Caja G, Castillo V, Albanell E, Such XJ: Response to lactation induction differs by season of year and breed of dairy ewes. Journal of Dairy Science. 2008; 91(6), 2299–2306. doi: 10.3168/jds.2007-0687.

[19] Misztal T, Górski K, Tomaszewska-Zaremba D, Molik E, Romanowicz K: Identification of salsolinol in the mediobasal hypothalamus of lactating ewes and its relation to suckling-induced prolactin and GH release. Journal of Endocrinology. 2008; 198, 83–89.

[20] Zieba D, Szczęsna M, Klocek-Górka B, Molik E, Misztal T, Williams GL, Romanowicz K, Stepien E, Keisler DH, Murawski M: Seasonal effects of central leptin infusion on melatonin and prolactin secretion and on SOCS-3 gene expression in ewes. Journal of Endocrinology. 2008; 198, 147–155.

[21] Górski K, Romanowicz K, Herman A, Molik E, Gajewska A, Tomaszewska-Zaremba D, Misztal T: The possible involvement of salsolinol and hypothalamic prolactin in the central regulatory processes in ewes during lactation. Reproduction in Domestic Animals. 2009;45(5), e54–e60. doi: 10.1111/j.1439-0531.2009.01521.

[22] Górski K, Romanowicz K, Molik E, Fülöp F, Misztal T: Effects of salsolinol and its antagonistic analogue, 1-MeDIQ, on growth hormone release in nursing sheep. Acta Neurobiology of Experimental. 2010; 70, 20–27.

[23] Reinert BD, Wilson FE: Hormone thyroid and the hypothalamus-pituitary-ovarian axis in American Tree Sparrows (Spizella arborea). General and Comparative Endocrinology. 1996; 103, 60–70.

[24] Karsch FJ, Dahl GE, Hachigian TM, Thrun LA: Involvement of thyroid hormones in seasonal reproduction. Journal of Reproduction and Fertility. Supplement. 1995; 49, 409–422.

[25] Klocek-Górka B, Szczęsna M, Molik E, Zieba D: The interaction of season, leptin and melatonin levels with thyroid hormone secretion, using an in vitro approach. Small Ruminant Research. 2010; 2, 231–235.

[26] Mitin V, Mikulec K, Karadjole I: Thyroid hormones and insulin concentration in sheep. Veterinarski Arhiv. 1986; 55, 73–75.

[27] Tucker HA: Hormones, mammary growth and lactation: a 41-year perspective. Journal of Dairy Science. 2000; 83, 874–884.

[28] Kokot F, Stupnicki R: Methods used radioimmunoassay and radiokompetycyjne in the clinic. PZWL, Warszawa, 1985.

[29] Morgan PJ: The pars tuberalis: the missing link in the photoperiodic regulation of prolactin secretion. Journal of Neuroendocrinology. 2000; 12, 287–295.

[30] Rowan W. Relation of light to bird migration and developmental changes. Nature. 1925; 115, 494–495.

Chapter 4

# Postlactational Involution: Molecular Mechanisms and Relevance for Breast Cancer Development

Edith C. Kordon and Omar A. Coso

Additional information is available at the end of the chapter

http://dx.doi.org/10.5772/66526

**Abstract**

Mammary gland tissue changes appearance and functionality in different sequential steps. The tissue of virgin, pregnant, or lactating mammary glands changes controlled by finely regulated physiological processes. A fourth stage (involution), triggered upon weaning, involves remodeling, and the gland regresses to resemble a prepregnant stage. This highly complex process characterized by a high degree of epithelial cell death and tissue remodeling can be divided into phases, which can be independent of each other. The present article describes a variety of signaling pathway components, transcription factors, and mRNA stabilization proteins that play a role in the regulation of cell fate during the involution process. These molecular actors are finely related in health to trigger the delicate mechanism that govern involution after weaning, leaving the gland in a latent stage until needed again. Importantly, it has been shown that this process may contribute to cancer development in the years following childbirth, mainly because of the involvement of inflammatory and remodeling factors.

**Keywords:** mammary gland involution, transcription factors, inflammatory cytokines, STATs, cell death, TNF-$\alpha$, mRNA stability, breast cancer

## 1. Introduction

Upon weaning, the mammary gland recovers a morphology similar to its prepregnant state through an intricate process known as postlactational involution. During this phase, a high proportion of epithelial cells die, the basal membrane is partially digested, and the adipose tissue reoccupies the space left by the regressed alveoli. In the mouse, mammary gland involution has been described as a two-step process according to its reversibility [1]. The first reversible phase is induced by *local factors* and lasts approximately 48 h during which it can be reversed through resuckling. At this stage, proapoptotic factors are upregulated, while survival factors are reduced. This reversible phase is followed after 72 h by a nonreversible phase where widespread apoptosis and tissue remodeling takes place. This later phase requires *systemic factors and local* proteases such as MMP-3/Stromelysin-1 and MMP11/Stromelysin-3, MMP2/Gelatinase A, ICE (interleukin-1 beta converting enzyme), and Urokinase-type plasminogen activator [2].

In 1997, Li et al. [3] defined the role of local factors as compared with systemic hormones during the first and second stages of involution. When milk release was disrupted in the presence of systemic lactogenic hormones, they demonstrated that local signals were sufficient to induce alveolar cell death. These authors demonstrated that a variety of procedures successfully triggered mammary involution, although none of them prevented the presence of circulating lactogenic hormones. For example, sealing of the teats, mammary gland transplants unable to release milk due to the absence of a teat connection or inactivation of the oxytocin gene efficiently induced mammary cell death. On the other hand, in these scenarios, systemic hormones were able to preserve lobular-alveolar structure, although they did not prevent apoptosis. This chapter reviews the discovery of the local factors involved in the process of involution as well as the finding of the mechanisms involved in their ability to induce cell death shortly after weaning. In addition, the generation of a pro-oncogenic microenvironment during mammary involution, which is able to facilitate breast cancer progression, is also discussed.

## 2. Involvement of STAT3 signaling

The studies referred above found that cell death correlated with the induction of proapoptotic genes as *bax*, decreased expression of milk proteins, dephosphorylation of STAT5a and 5b (main transcription factors that mediate prolactin triggered signaling), and activation of STAT3. Signal transducer and activator of transcription (STATs) are a family of latent transcription factors, which are activated in response to a variety of cytokines and growth factors. This family of signaling molecules has been implicated in cell growth, differentiation, survival, and apoptosis. STAT3 and STAT5 have reciprocal patterns of activation throughout a mammary developmental cycle, suggesting that STAT5 may be a survival factor and STAT3 a death factor for differentiated mammary epithelium. Chapman et al. [4], using the lox/Cre recombination system, showed a decrease in epithelial apoptosis and a dramatic delay of the involution process upon forced weaning in conditional KO mice, in which STAT3 was specifically deleted in the lactating mammary gland. In addition, early activation of STAT1 and induction of p53 and p21 expression was observed, which suggested a potential compensatory mechanism for induction of eventual involution in the STAT3 null mammary glands. These results demonstrated the importance of STAT factors in signaling the initiation of physiological apoptosis *in vivo* and highlighted the utility of the lox/Cre system for addressing the function of genes with an embryonic lethal phenotype, specifically in the mammary gland.

STAT3 is the most ubiquitous of the members of this family of proteins, is activated by many different cytokines and growth factors, and plays many roles in different physiological processes. In addition, this protein is, the first STAT family member found to be constitutively activated in a variety of neoplastic tissues. It was determined that STAT3 modulates the expression of various target genes involved in cell-cycle regulation, angiogenesis, and apoptosis inhibition. Because of this, silencing or inhibiting STAT3 reduces tumor cell proliferation and survival in both animal and human studies. However, in the involuting mammary gland, STAT3 signaling induces cell death [5].

By 2003, it had been shown that STAT3 is the main factor involved in the initiation of apoptosis of mammary cells after weaning, but the mechanism of its activation remained unclear. In 2002, based on the hypothesis that IL-6 is the activating cytokine for STAT3, Hennighausen's group showed that expression of IL-6 increases during early involution together with STAT3 and p44/42 MAPK activation. Besides, it was shown that IL-6 treatment activated STAT3 in the mammary gland of virgin and lactating mice. In addition, IL-6-, STAT3-, and Bax-null mice showed similar mammary phenotypes, that is a significant delay in postlactational involution. Nevertheless, it was demonstrated that STAT3 activation during involution was independent of the IL-6 levels in the mammary after weaning. In contrast, the increase of p44/42 MAPK (ERK1/2) phosphorylation at the onset of involution was dependent on the presence of this cytokine. This suggested that either IL-6 does not induce STAT3 *in vivo* or its absence is compensated for by other cytokines, such as leukemia-inhibitory factor (LIF) [6].

By that time, there was no evidence in the literature reporting LIF expression and/or activities in the normal mammary gland tissue. Therefore, LIF expression profile was analyzed during the successive stages of mammary gland development, function, and involution. The results demonstrated that LIF is expressed in the mammary gland at low levels in postpubertal, adult virgin, and pregnant mice. But, expression of this protein almost disappear during lactation to then show a significant increase a few hours after weaning, maintaining these high levels during the following days. We demonstrated that LIF expression in the gland is induced by milk stasis and not by the decrease of circulating lactogenic hormones after weaning. In addition, implantation of LIF containing pellets in lactating glands resulted in a significant increase of STAT3 phosphorylation and epithelium apoptosis. We then concluded that LIF-regulated expression in the mouse mammary gland may play a relevant role during the first stage of mammary gland involution and that LIF-induced mammary epithelium apoptosis could be mediated, at least partially, by STAT3 activation [7].

Shortly after our paper was published, Christine Watson's lab also demonstrated that LIF is the physiological activator of STAT3, as they report that pSTAT3 is absent and C/EBPdelta (a well-known STAT3 target) is not upregulated in involuting transplanted mammary glands of LIF double knock-out (LIF(−/−) mice). Similarly to what was observed in the STAT3-null glands, LIF(−/−) mammary glands exhibit delayed involution, reduced apoptosis, and elevated levels of p53 [8].

STAT3 activation and LIF expression have not been observed only in the involuting mammary gland. It was determined that autocrine/paracrine LIF present in conditioned medium from primary cultures of mouse mammary tumors was also able to induce activation of that transcription factor and to increase cell survival in mammary tumor cell lines. However, although LIF blocking antibody prevented STAT3 phosphorylation, inhibition of STAT3 increased cell survival. These results indicated that LIF is overexpressed in mouse mammary tumors, where it acts as the main STAT3 activator. Nevertheless, the data also suggested that the positive

LIF effect on tumor cell survival was not dependent on STAT3 activation, which seemed to inhibit tumor cell viability as it does in involuting mammary epithelium [9]. Kritikou et al. showed that pERK1/2 is significantly reduced in LIF(−/−) glands during pregnancy [8], suggesting that at this stage, LIF mediates its effects through pERK1/2. Therefore, it is possible that LIF proliferative effects on mammary tumors depend on ERK ½ activation. In addition, although it has been reported that STAT3 acts a potent oncogene in different tumor types, it was also demonstrated that the biological role of this factor is modulated by the stage of tumor progression [10]. Similarly, it can be proposed that in well-to-moderately differentiated mammary tumors, STAT3 activation induces cell death as observed in nontumorigenic mammary cells after lactation. This activity might be altered in more aggressive or less differentiated tumors, as it has been shown that STAT3 constitutive activation is very common in basal breast cancer [11], which have worse prognosis than luminal tumors. However, our results imply that in the development of therapeutic strategies for blocking STAT3 in breast cancer cells, the strong dependence on the cellular context that this factor activity displays should be taken into account.

Mechanical stress is a relevant factor to induce adaptive responses in multiple cell types [12–16]. Importantly, the signaling pathways triggered by this stimulus in those different examples also play a relevant role during mammary gland involution. Therefore, it was proposed that upon weaning, milk accumulation may cause cell stretching that, in turn, would induce the initiation of the molecular cascades that lead to the remodeling process of the lactating gland. To address this issue, we designed a new practical device that allowed us to evaluate the effects of radial stretching on the HC11 nontumorigenic mammary epithelial cell line cultured on flexible silicone membranes. The results showed that, as previously observed in other cell types, mechanical stress induced ERK1/2 phosphorylation and c-Fos expression induction, as well as LIF secretion, STAT3 activation, and AKT phosphorylation inhibition. Therefore, mechanical strain is able to induce weaning-associated events in cultured mammary epithelial cells [17].

STAT3 is essential, but not sufficient for the onset of apoptosis during mammary involution, as expression of a constitutively active *Akt*, a downstream effector of the phosphoinositide-3-OH kinase (PI3K) pathway, provides an overriding survival signal after lactation [18]. However, AKT downregulation depends on STAT3 activation, since PI(3)K regulatory subunits p55$\alpha$ and p50$\alpha$ (each of them, when overexpressed, reduces levels of activated AKT) are induced by that transcription factor during mammary involution. In fact, it has been shown that STAT3 binds directly to the promoters of p55$\alpha$ and p50$\alpha$ subunits *in vivo* and in STAT3 KO mice, upregulation of p55$\alpha$ and p50$\alpha$ is abrogated, levels of activated AKT are sustained, and apoptosis is prevented [19]. In addition, it was shown that deletion of both p55$\alpha$ and p50$\alpha$ subunits reduced cell death as well as expression and activity of cathepsin L during mammary involution. This protease participates in lysosomal-mediated programmed cell death (LM-PCD), which is upregulated during normal involution by activated STAT3. Furthermore, involution is delayed in cathepsin L-deficient mice, suggesting that the p55$\alpha$/p50$\alpha$ subunits mediate cell death in part by elevating the level of cathepsin L. Surprisingly, it was found that during involution, p55$\alpha$/p50$\alpha$ localize to the nucleus where they bind to chromatin and regulate transcription of a subset of inflammatory/acute phase genes that are also STAT3 targets. Therefore, these findings revealed that postlactational regression of the mammary gland is accomplished through a nonclassical, lysosomal-mediated pathway of cell death, in which PI3K regulatory subunits participate as main regulators [20]. In fact, it has been demonstrated that cell death of mammary epithelium after weaning does not depend on the activation of executioner caspases 3, 6, and 7, although it requires STAT3 for cathepsin B and L induction as well as for the downregulation of their endogenous inhibitor Spi2A [21].

Global gene expression changes during involution have been profiled by microarray analysis, which allowed characterization of clusters of genes with distinct expression profiles during the first 4 days of involution. Such expression profiling led to the observation that one of the most strikingly upregulated genes in the absence of STAT3 is the serpin Spi2a. Interestingly, during mammary involution, STAT3 not only regulates LM-PCD by inhibiting serpin Spi2a, inducing the expression of cathepsins B and L, and the regulatory subunits p55$\alpha$/p50$\alpha$, but also by the uptake of secreted MFGs that lead to the formation and fusion of large lysosomal-like vacuoles, which are toxic to epithelial cells. Upon re-entry, the MFG (mammary fat globules) triglycerides are metabolized to free fatty acids, including oleic acid, that can distort membranes and result in leakage of cathepsins from lysosomes. Therefore, STAT3 promotes a phenotypic switch from secretion to phagocytosis of MFGs, the latter function delivering triglyceride to vacuoles with the ensuing consequences of LMP and cell death [22].

## 3. NF-κB signaling

It is clear that there are multiple mechanisms of regulation in early involution that synergise to ensure efficient induction of cell death, phagocytosis, suppression of inflammation, and remodeling of the architecture of the gland. Clarkson and Watson identified clusters of genes that are transcriptional targets of either NF-κB or STAT3, or indeed both, during early involution [23]. For example, among the NFκB targets, the TNF superfamily of death receptor (DR) ligands have been detected. These proteins induce apoptosis through binding to their receptor, which recruits caspase 8 (via FADD) and activates executioner caspases, finally leading to cell death [24]. Specifically, *Tnf, Tnfsf4, Tnfsf6, Tnfsf7, Tnfsf10,* and *Tnfsf12* are induced transiently at 12 h after weaning, and the proteins Fas ligand, TNF-$\alpha$, TWEAK, and TRAIL are able to activate extrinsic apoptosis through their cognate receptors Fas, TNFR-1, TNFR-2, DR3, and DR4. The genes for the first two of these receptors (*Fas* and *Tnfrsf1a*) were also induced, and maximally coexpressed, within 24 h of weaning. NF-κB activity also correlated with the rapid activation of these TNF superfamily ligands [25].

Particularly, about TNF-$\alpha$, our results have shown that this factor, through TNF-$\alpha$ receptor-2 (TNFR2) binding induces LIF expression mediated by ERK1/2 activation in nontumorigenic mouse mammary epithelial cells. In addition, the AP-1 has been implicated in this signaling cascade, since blocking the activity of this transcription factor resulted in a significant reduction of TNF-$\alpha$ induced LIF expression. Therefore, TNF-$\alpha$ may contribute to mammary gland involution by, among other activities, eliciting LIF expression through ERK1/2 and AP1 activation [26].

The NF-κB family of transcription factors primarily plays anti-apoptotic roles. DNA binding activity of this transcription factor is markedly upregulated within 3 h of forced involution and is suggested to promote survival of a subpopulation of mammary epithelial cells [25]. This hypothesis is consistent with the paradigm of NF-κB–mediated suppression of TNF-α cytotoxicity in TNF-α–responsive cells. NF-κB activity is mediated by a multiprotein signaling complex called the IκB kinase (IKK), which consists of two catalytic subunits: IKK1/α, IKK2/β and a regulatory subunit, NEMO (NF-kappa-B essential modulator). Activation of this complex leads to phosphorylation of the IκB proteins; phospho-IκB is rapidly ubiquitinated and degraded via the 26S-proteasome releasing NF-κB and unmasking its nuclear localization signal, allowing its activity as transcription regulator of many target genes [27]. NF-κB then inhibits the death signal by trans-activating genes that promote resistance to apoptosis. The effect of this negative feedback mediated by NF-κB is the modulation of apoptosis in response to the TNF-α death signal. However, deletion of the gene encoding IKK2 resulted in delayed apoptosis and remodeling, as well as blockade of caspase 3 activation in the postlactational mammary gland. This failure to induce cell death was associated with reduced expression of TNF and its receptor TNFR1, which are known NF-κB targets. In addition, the observed high levels of active AKT together with downregulation of TWEAK, another DR ligand, also contributed to retard the involution process in these genetically engineered mice [28]. These results suggest that NF-κB may provide either proapoptotic or antiapoptotic signals during involution, depending on the timing and cellular context in which this transcription factor is activated.

## 4. Gene expression regulation at the level of mRNA stability

It has been demonstrated that the stability of many messenger RNAs (mRNAs) encoding oncoproteins, chemokines, cytokines, and other inflammatory mediators is controlled by AU-rich elements (AREs), sequences located within the 3′-UTR of many transcripts [29, 30]. ARE-directed control of mRNA decay is mediated, in part, through interactions with specific ARE-binding proteins (AUBPs). One such protein is tristetraprolin (TTP), which accelerates the decay of targeted transcripts [31]. During inflammation, TTP plays a relevant role destabilizing different mRNAs, participating in glucocorticoid-mediated anti-inflammatory activity [32–34] and inhibiting NF-κB signaling [35]. The relevance of TTP as a negative regulator of these processes has been demonstrated by the severe chronic inflammation displayed by multiple tissues in TTP-KO mice, which was mostly due to the dramatic increase of TNF-α levels [32].

Several reports indicate that TTP participates in the inhibition of tumor progression. It has been shown that TTP mRNA levels are significantly decreased in many tumor types, including breast cancer [36]. We have also reported that TTP expression is lower in all breast cancer types compared with normal mammary tissue, and high levels of this protein negatively correlate with cancer cell aggressiveness. Interestingly, we have also determined that in the mouse mammary gland, expression of this protein reaches the highest level during lactation, and can be induced in culture by treatment with lactogenic hormones [37].

http://dx.doi.org/10.5772/66526

These studies reveal a new potential biological role for this tumor suppressor protein in mammary epithelium, since TTP might protect the tissue from inflammatory and/or remodeling activities that would trigger involution of the gland. Interestingly, our unpublished results show that by reducing TTP expression in the differentiated mammary epithelium, cell death is induced in the midst of lactation without requirement of additional stimuli. Then, TTP is not (or at least not only) a mechanism of surveillance, which prevents an eventual increase of inflammatory factors that might lead lactation to a halt, but it actually functions as a survival factor in the mammary epithelium, since reducing its levels is enough to induce cell death and involution of this tissue.

## 5. Mammary involution and cancer

Breast cancer is the most frequent malignancy diagnosed in association with pregnancy [38–40]. In addition, different studies have demonstrated an increase in breast cancer risk in the years immediately following giving birth [41–43]. Importantly, not only a rise in breast cancer incidence during the postpartum years has been observed, but also a higher risk for poor outcomes in women diagnosed during that timeframe [44]. However, it has been well established that pregnancy provides lifetime protection for women who are under 35 years at first birth [45–47]. Therefore, pregnancy would exert two opposite effects on breast cancer development: induces protection, which is associated with the differentiation of mammary epithelium, and increases risk, through alteration of tissue microenvironment.

Postpartum breast cancers have been referred to as type II Pregnancy-Associated Breast Cancer (PABC) to distinguish these cancers from those diagnosed during pregnancy [45]. A physiological window unique to type II PABC is mammary gland involution and, in an effort to distinguish why type II PABC patients have worse prognoses, the normal postpartum breast microenvironment has been investigated for potential tumor-enhancing attributes. These studies reveal that postpartum involution utilizes wound-healing programs for gland remodeling, including increases in matrix metalloproteinase activity, release of bioactive fragments of extracellular matrix (ECM) termed matricryptins, accumulation of fibrillar collagen, and influx of immune cells, which also generates tumor-promotional microenvironments [48–51]. *In vivo* experiments showed that tumors from xenografts of breast cancer cells exposed to the postpartum involution microenvironment have increased growth, invasion, and metastasis compared to those growing in nulliparous hosts. In addition, these implants showed augmented fibrillar collagen accumulation and high cyclooxygenase-2 (COX-2) expression [52]. Coincidently, in the postlactating mammary gland, inhibition of COX-2 reduced the collagen fibrillogenesis, as well as tumor development and cancer cell invasiveness [53]. Therefore, it has been proposed that women at high risk for postpartum breast cancer might benefit from treatment with nonsteroidal anti-inflammatory drugs (NSAIDs) during postpartum, which would reduce COX-2 expression and its consequences on the behavior of breast initiated cells.

Without doubt, immune response plays a primordial role in mouse mammary involution, since molecular profiling of that phase is consistent with acute phase, innate and adaptive immune responses [51–54]. Interestingly, after weaning, mammary epithelial cells themselves express transcripts traditionally associated with immune cells [55, 56] and acquire phagocytic capability [57]. Therefore, it has not been possible to completely determine which cell types and in what part are responsible for the observed immune-like gene signatures. Particularly, there is not much data about the participation of adaptive immune cells, but innate immune cell populations have been partially characterized. Specifically, it has been observed that granulocyte infiltration in the mouse gland on the first day of involution suggest the involvement of this cell type in early involution [58]. In addition, resident macrophages seem to be required for this phase, while infiltrating macrophages are important during the remodeling stage [52]. It was observed that on this last phase, macrophages express low iNOS, high arginase-1, and the mannose receptor, which is consistent with alternative activation or M2 polarization of these cells [59]. Importantly, this phenotype correlated with breast tumor promotion in patients [60] and murine mammary tumor progression [61].

The earlier mentioned STAT3 and NF-κB signaling pathways, as well as others involving transforming growth factor beta (TGF-β) and the retinoid acid receptors (RARs)/retinoid X receptors (RXRs), participate in mammary gland involution as well as in breast cancer development. After weaning, target genes of RARα/p300 and RelA/p65, which belong to the NF-κB protein family, are induced, and high activity of the proteins coded by these genes, e.g., *MMP9*, *Capn1*, and *Capn2*, has been detected in breast cancer cells. Calpains belong to a family of calcium-dependent intracellular cysteine proteases involved in a wide variety of physiological and pathological processes. These proteases are heterodimers, consisting of a small regulatory subunit, encoded by CAPN4 gene, common for both members, and a large catalytic subunit encoded by either CAPN1 or CAPN2. During mammary gland involution and cancer progression, these proteins are relevant for modifying the extracellular matrix, allowing tissue remodeling and/or cell invasion. In addition, calpains also cleave intracellular proteins located in the cell membrane, lysosomes, mitochondria, and nuclei, favoring cell death during involution and cell anchoring loss during tumor progression [62].

## 6. Conclusion

As in other physiological processes, the use of conditional knockout mice and the application of high-throughput techniques have been very useful to understand that normal postlactation mammary gland involution relays on the fine-tune coordination of multiple signaling pathways. Although involution is a physiologically normal, developmentally orchestrated tissue-remodeling process, it shares striking similarities with pathologically induced wound-healing and tumor-promotional microenvironments. This highlights the relevance of further investigating this process, since it may yield novel therapeutic targets or prognostic markers for breast cancer. Importantly, studying this particular phase of mammary gland biology may help us to pinpoint subtle changes in a pro-oncogenic environment that determines the derailment from normal physiological to pathological tissue behavior.

## Author details

Edith C. Kordon[1]* and Omar A. Coso[2]

*Address all correspondence to: ekordon@gmail.com

1 Department of Biological Chemistry, Institute of Physiology, Molecular Biology and Neuroscience (IFIBYNE-UBA-CONICET), School of Exact and Natural Sciences, University of Buenos Aires, (FCEN-UBA), Ciudad Universitaria, Buenos Aires, Argentina

2 Department of Physiology, Molecular and Cellular Biology, Institute of Physiology, Molecular Biology and Neuroscience (IFIBYNE-UBA-CONICET), School of Exact and Natural Sciences, University of Buenos Aires (FCEN-UBA), Ciudad Universitaria, Buenos Aires, Argentina

## References

[1] Lund LR, Romer J, Thomasset N, Solberg H, Pyke C, Bissell MJ, et al. Two distinct phases of apoptosis in mammary gland involution: proteinase-independent and -dependent pathways. Development. 1996;**122**:181–193.

[2] Green KA, Lund LR. ECM degrading proteases and tissue remodelling in the mammary gland. Bioessays. 2005;**27**:894–903.

[3] Li M, Liu X, Robinson G, Bar-peled U, Wagner K, Young S, Hennighausen L, and Furth P. Mammary-derived signals activate programmed cell death during the first stage of mammary gland involution. Proc Natl Acad Sci U S A. 1997;**94**:3425–3430

[4] Chapman RS, Lourenco PC, Tonner E, Flint DJ, Selbert S, Takeda K, Akira S, Clarke AR, Watson CJ. Suppression of epithelial apoptosis and delayed mammary gland involution in mice with a conditional knockout of Stat3. Genes Dev. 1999;**13(19)**:2604–2616.

[5] Resemann HK, Watson CJ, Lloyd-Lewis B. The STAT3 paradox: a killer and an oncogene. Mol Cell Endocrinol. 2014;**382(1)**:603–611.

[6] Zhao L, Melenhorst JJ, Hennighausen L. Loss of interleukin 6 results in delayed mammary gland involution: a possible role for mitogen-activated protein kinase and not signal transducer and activator of transcription 3. Mol Endocrinol. 2002;**12**:2902–2912.

[7] Schere-Levy C, Buggiano V, Quaglino A, Gattelli A, Cirio MC, Piazzon I, Vanzulli S, Kordon EC. Leukemia inhibitory factor induces apoptosis of the mammary epithelial cells and participates in mouse mammary gland involution. Exp Cell Res. 2003;**282(1)**:35–44.

[8] Kritikou EA, Sharkey A, Abell K, Came PJ, Anderson E, Clarkson RW, Watson CJ. A dual, non-redundant, role for LIF as a regulator of development and STAT3-mediated cell death in mammary gland. Development. 2003;**130**(**15**):3459–3468.

[9] Quaglino A, Schere-Levy C, Romorini L, Meiss RP, Kordon EC. Mouse mammary tumors display STAT3 activation dependent on leukemia inhibitory factor signaling. Breast Cancer Res. 2007;**9**(**5**):R69.

[10] Lu C, Kerbel RS. Interleukin-6 undergoes transition from paracrine growth inhibitor to autocrine stimulator during human melanoma progression. J Cell Biol. 1993;**120**:1281–1288.

[11] Yue P and Turkson J. Targeting STAT3 in cancer: how successful are we?. Expert Opin Investig Drugs. 2009;**18**(**1**):45–56.

[12] Trepat X, Puig F, Gavara N, Fredberg JJ, Farre R, Navajas D. Effect of stretch on structural integrity and micromechanics of human alveolar epithelial cell monolayers exposed to thrombin. Am J Physiol Lung Cell Mol Physiol. 2006;**290**:L1104–L1110.

[13] Ning QM, Wang XR. Response of alveolar type II epithelial cells to mechanical stretch and lipopolysaccharide. Respiration. 2007;**74**:579–585.

[14] Wilson E, Sudhir K, Ives HE. Mechanical strain of rat vascular smooth muscle cells is sensed by specific extracellular matrix/integrin interactions. J Clin Invest. 1995;**96**:2364–2372.

[15] Perrone CE, Fenwick-Smith D, Vandenburgh HH. Collagen and stretch modulate autocrine secretion of insulin-like growth factor-1 and insulin-like growth factor binding proteins from differentiated skeletal muscle cells. J Biol Chem. 1995;**270**:2099–2106.

[16] Sadoshima J, Xu Y, Slayter HS, Izumo S. Autocrine release of angiotensin II mediates stretch-induced hypertrophy of cardiac myocytes in vitro. Cell. 1993;**75**:977–984.

[17] Quaglino A, Salierno M, Pellegrotti J, Rubinstein N, Kordon EC. Mechanical strain induces involution-associated events in mammary epithelial cells. BMC Cell Biol. 2009;**14**:1471–2121.

[18] Schwertfeger KL, Richert MM, Anderson SM. Mammary gland involution is delayed by activated Akt in transgenic mice. Mol Endocrinol. 2001,**15**(**6**):867–881.

[19] Abell K, Bilancio A, Clarkson RW, Tiffen PG, Altaparmakov AI, Burdon TG, Asano T, Vanhaesebroeck B, Watson CJ. STAT3-induced apoptosis requires a molecular switch in PI(3)K subunit composition. Nat Cell Biol. 2005;**4**:392–398.

[20] Pensa S, Neoh K, Resemann HK, Kreuzaler PA, Abell K, Clarke NJ, Reinheckel T, Kahn CR, Watson CJ. The PI3K regulatory subunits p55$\alpha$ and p50$\alpha$ regulate cell death in vivo. Cell Death Differ. 2014;**21**(**9**):1442–1450.

[21] Kreuzaler PA, Staniszewska AD, Li W, Omidvar N, Kedjouar B, Turkson J, Poli V, Flavell RA, Clarkson RW, Watson CJ. STAT3 controls lysosomal-mediated cell death in vivo. Nat Cell Biol. 2011;**13**(**3**):303–309.

[22] Sargeant TJ, Lloyd-Lewis B, Resemann HK, Ramos-Montoya A, Skepper J, Watson CJ. STAT3 controls cell death during mammary gland involution by regulating uptake of milk fat globules and lysosomal membrane permeabilization. Nat Cell Biol. 2014;**16**(**11**):1057–1068.

[23] Clarkson RW, Watson CJ. Microarray analysis of the involution switch. J Mammary Gland Biol Neoplasia. 2003;**8**(**3**):309–319.

[24] Ashkenazi A, Dixit VM. Death receptors: signaling and modulation. Science. 1998;**281**(**5381**):1305–1308.

[25] Clarkson RW, Heeley JL, Chapman R, Aillet F, Hay RT, Wyllie A, Watson CJ. NF-kappaB inhibits apoptosis in murine mammary epithelia. J Biol Chem. 2000;**275**(**17**):12737–12742.

[26] Levy CS, Slomiansky V., Gattelli A., Nahmod K., Pelisch F., Blaustein M., Srebrow A., Coso O. and Kordon E. Tumor necrosis factor alpha induces LIF expression through ERK1/2 activation in mammary epithelial cells. J Cell Biochem. 2010;**110**:857–865.

[27] Karin, M and Ben-Neriah, Y. Phosphorylation meets ubiquitination: the control of NF-kB activity. Annu Rev Immunol. 2000;**18**:621–663.

[28] Baxter FO, Came PJ, Abell K, Kedjouar B, Huth M, Rajewsky K, Pasparakis M, Watson CJ. IKK/2 induces TWEAK and apoptosis in mammary epithelial cells. Development. 2006;**133**:3385–3394.

[29] Chen CYA, Shyu AB. AU-rich elements: characterization and importance in mRNA degradation. Trends Biochem Sci 1995;**20**:465–470.

[30] Guhaniyogi J, Brewer G. Regulation of mRNA stability in mammalian cells. Gene. 2001;**265**:11–23.

[31] Barreau C, Paillard L, Osborne HB. AU-rich elements and associated factors: are there unifying principles?. Nucleic Acids Res. 2005;**33**:7138–7150.

[32] Taylor GA, Carballo E, Lee DM, Lai WS, Thompson MJ, Patel D, Schenkman DI, Gilkeson GS, Broxmeyer HE, Haynes BF, Blackshear PJ. A pathogenetic role for TNF alpha in the syndrome of cachexia, arthritis, and autoimmunity resulting from tristetraprolin (TTP) deficiency. Immunity. 1996; May;4(**5**):445–454.

[33] Smoak K, Cidlowski JA. Glucocorticoids regulate triste-traprolin synthesis and posttranscriptionally regulate tumor necrosis factor alpha inflammatory signaling. Mol Cell Biol. 2006;**26**:9126–9135.

[34] Ishmael FT, Fang X, Galdiero MR, Atasoy U, Rigby WF, Gorospe M, Cheadle C, Stellato C. Role of the RNA-binding protein tristetraprolin in glucocorticoid-mediated gene regulation. J Immunol. 2008;**180**:8342–8353.

[35] Schichl YM, Resch U, Hofer-Warbinek R, de Martin R. Tristetraprolin impairs NF-B/p65 nuclear translocation. J Biol Chem. 2009;**284**(**43**):29571–29581.

[36] Brennan S, Kuwano Y, Alkharouf N, Blackshear PJ, Gorospe M, Wilson GM. The mRNA-destabilizing protein tristetraprolin is suppressed in many cancers, altering tumorigenic phenotypes and patient prognosis. Cancer Res. 2009;**69**:5168–5176.

[37] Goddio M, Gattelli A, Slomiansky V, Lacunza E, Gingerich T, Curino A, Facchinetti M, LaMarre J, Abba M, Kordon E. Mammary differentiation induces expression of Tristetraprolin, a tumor suppressor AU-rich mRNA-binding protein. Breast Cancer Res Treat. 2012;**135**(**3**):749–758.

[38] Rodriguez AO, Chew H, Cress R, Xing G, McElvy S, Danielsen B, Smith L. Evidence of poorer survival in pregnancy associated breast cancer. Obstet Gynecol. 2008;**112**(**1**):71–78.

[39] Cardonick E, Dougherty R, Grana G, Gilmandyar D, Ghaffar S, Usmani A. Breast cancer during pregnancy: maternal and fetal outcomes. Cancer J. 2010;**16**(**1**):76–82.

[40] Pregnancy and Breast Cancer 2011; Available from: http://www.cancer.org/Cancer/BreastCancer/MoreInformation/pregnancy-andbreast-cancer. Accessed 04 October 2016.

[41] Lambe M, Hsieh C, Trichopoulos D, Ekbom A, Pavia M, Adami HO. Transient increase in the risk of breast cancer after giving birth. N Engl J Med. 1994;**331**(**1**):5–9.

[42] Lord SJ, Bernstein L, Johnson KA, Malone KE, McDonald JA, Marchbanks PA, Simon MS, Strom BL, Press MF, Folger SG, Burkman RT, Deapen D, Spirtas R, Ursin G. Breast cancer risk and hormone receptor status in older women by parity, age of first birth, and breastfeeding: a case-control study. Cancer Epidemiol Biomarkers Prev. 2008;**17**(**7**):1723–1730.

[43] Liu Q, Wuu J, Lambe M, Hsieh SF, Ekbom A, Hsieh CC. Transient increase in breast cancer risk after giving birth: postpartum period with the highest risk (Sweden). Cancer Causes Control. 2002;**13**(**4**):299–305.

[44] Callihan EB, Gao D, Jindal S, Lyons TR, Manthey E, Edgerton S, Urquhart A, Schedin P, Borges VF. Postpartum diagnosis demonstrates a high risk for metastasis and merits an expanded definition of pregnancy-associated breast cancer. Breast Cancer Res Treat. 2013;**138**:549–559.

[45] Lyons TR, Schedin PJ, Borges VF. Pregnancy and breast cancer: when they collide. J Mammary Gland Biol Neoplasia. 2009;**14**(**2**):87–98.

[46] Rosner B, Colditz GA. Nurses' health study: log-incidence mathematical model of breast cancer incidence. J Natl Cancer Inst. 1996;**88**(**6**):359–364.

[47] Schedin P. Pregnancy-associated breast cancer and metastasis. Nat Rev Cancer. 2006;**6**(**4**):281–291.

[48] Lund L, Romer J, Thomasset N, Solberg H, Pyke C, Bissell M, Dano K, Werb Z. Two distinct phases of apoptosis in mammary gland involution: proteinase-independent and -dependent pathways. Development (Cambridge, England). 1996;**122**:181–193.

[49] Schedin P, Strange R, Mitrenga T, Wolfe P, Kaeck M. Fibronectin fragments induce MMP activity in mouse mammary epithelial cells: evidence for a role in mammary tissue remodeling. J Cell Sci. 2000;**113** (Pt 5):795–806.

[50] Schedin P, Mitrenga T, Mcdaniel S, Kaeck M. Mammary ECM composition and function are altered by reproductive state. Molec Carcinog. 2004;**41**:207–220.

[51] O'Brien J, Lyons T, Monks J, Lucia M, Wilson R, Hines L, Man Y, Borges V, Schedin P. Alternatively activated macrophages and collagen remodeling characterize the postpartum involuting mammary gland across species. Am J Pathol. 2010;**176**:1241–1255.

[52] Lyons T, O'Brien J, Borges V, Conklin M, Keely P, Eliceiri K, Marusyk A, Tan A, Schedin P. Postpartum mammary gland involution drives progression of ductal carcinoma in situ through collagen and COX-2. Nature Med. 2011;**17**:1109–1115.

[53] O'Brien J, Hansen K, Barkan D, Green J And Schedin P. Non-steroidal anti-inflammatory drugs target the pro-tumorigenic extracellular matrix of the postpartum mammary gland. Int J Dev Biol. 2011;**55**:745–755.

[54] Clarkson RW, Wayland MT, Lee J, Freeman T, Watson CJ. Gene expression profiling of mammary gland development reveals putative roles for death receptors and immune mediators in post-lactational regression. Breast Cancer Res 2004;**6**:R92–R109.

[55] Hughes K, Wickenden JA, Allen JE, Watson CJ. Conditional deletion of STAT3 in mammary epithelium impairs the acute phase response and modulates immune cell numbers during post-lactational regression. J Pathol. 2012;**227**:106–117.

[56] Nguyen AV, Pollard JW. Transforming growth factor beta3 induces cell death during the first stage of mammary gland involution. Development. 2000;**127**:3107–3118.

[57] Monks J, Smith-Steinhart C, Kruk ER, Fadok VA, Henson PM. Epithelial cells remove apoptotic epithelial cells during post-lactation involution of the mouse mammary gland. Biol Reprod. 2008;**78**:586–594.

[58] Stein T, Morris JS, Davies CR, Weber Hall SJ, Duffy MA, Heath VJ, Bell AK, Ferrier RK, Sandilands GP, Gusterson BA. Involution of the mouse mammary gland is associated with an immune cascade and an acute-phase response, involving LBP, CD14 and STAT3. Breast Cancer Res. 2004;**6**:R75–R91.

[59] Mantovani A, Sica A, Sozzani S, Allavena P, Vecchi A, Locati M. The chemokine system in diverse forms of macrophage activation and polarization. Trends Immunol. 2004;**25**:677–686.

[60] Lewis CE, Pollard JW. Distinct role of macrophages in different tumor microenvironments. Cancer Res. 2006;**66**:605–612.

[61] Qian BZ, Pollard JW. Macrophage diversity enhances tumor progression and metastasis. Cell. 2010;**141**:39–51.

[62] Zaragoza R, García-Trevijano ER, Lluch A, Ribas G, Viña JR. Involvement of different networks in mammary gland involution after the pregnancy/lactation cycle: implications in breast cancer. IUBMB Life. 2015;**67**(**4**):227–238.

Chapter 5

# Non-Coding RNA Roles in Ruminant Mammary Gland Development and Lactation

Duy N. Do and Eveline M. Ibeagha-Awemu

Additional information is available at the end of the chapter

http://dx.doi.org/10.5772/67194

**Abstract**

The ruminant mammary gland (MG) is an important organ charged with the production of milk for young and human nourishment. Many factors influence MG productivity, including nutrition, genetics, breed, epigenetics (including non-coding RNA [ncRNA]), disease pathogens and other environmental factors. In recent years, increasing research is beginning to determine the role of non-coding RNA in MG functions. Non-coding RNAs (small interfering RNA [siRNA], microRNA [miRNA], PIWI-interacting RNA [piRNA], small nucleolar RNA [snoRNA] and long non-coding RNA [lncRNA]) are a class of untranslated RNA molecules that function to regulate gene expression, associated biochemical pathways and cellular functions and are involved in many biological processes. This chapter presents a review of the current state of knowledge on the role of ncRNAs (particularly miRNAs and lncRNAs) in the MG and lactation processes, lactation signalling pathways, lipid metabolism, MG health of ruminants as well as miRNA roles in milk recipients. Finally, the potential application of new genome editing technology for ncRNA studies in MG development, the lactation process and milk components is presented.

**Keywords:** non-coding RNA, microRNA, long non-coding RNA, mammary gland, lactation, genome editing, signalling pathways

## 1. Introduction

As one of the remarkable products of evolution, lactation is a very dynamic and complex process. The process of lactation involves the development of the mammary gland (MG) and the synthesis and secretion of milk. The lactation process is affected by many factors, including genetics, epigenetics, non-genetics and environmental factors. The knowledge of lactation regulation is not only important for improvement of milk production and quality but also provides a model for basic cellular processes (proliferation, differentiation, survival and death) [1], which may have important implications for productivity (milk yield) and disease status (e.g. breast cancer, mastitis, etc.). The endocrine regulation and physiological processes as well as the signalling pathways involved in these processes are fairly understood [1, 2]. Facilitated by the release of the whole genome sequences of cattle, sheep and goat [3–6] as well as availability of single nucleotide polymorphism (SNP) genotyping chips [7–11], the genetic mechanisms of ruminant lactation have been extensively explored (**Figure 1**). As a consequence, many quantitative trait loci (QTL) and genetic markers for lactation-related traits (for instance, milk yield, milk components, lactation persistency, etc.) have been detected and catalogued in the animal QTL database (http://www.animalgenome.org/cgi-bin/QTLdb/index).

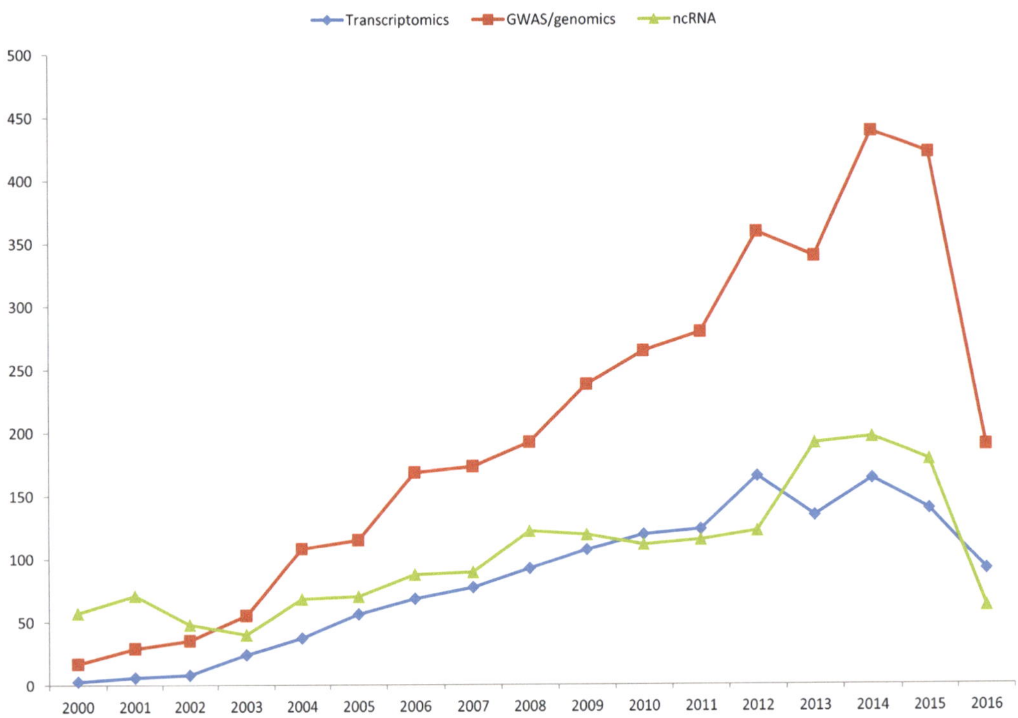


**Figure 1.** Growing research by year in the field of cattle genomics and transcriptomics (including non-coding RNA) from January 2000 to August 2016.

Transcriptomics research by both microarray and RNA sequencing methods has allowed for a better understanding of the genes and regulatory networks of complex traits in animals [12], such as the biosynthesis of major milk components (reviewed in Refs. [13, 14]). Emerging studies now suggest that non-coding RNAs (ncRNAs) are key regulators of mammary gland development and lactation processes [15–17]. The results from the ENCODE (ENCyclopedia of DNA elements) project [18, 19] indicate that only a small portion of the genome, about

1.5%, codes for proteins while most of the genome is transcribed into non-coding regulatory elements or ncRNA. This indicates that ncRNAs play significant regulatory roles in complex animal traits. A similar project to functionally annotate regulatory elements in animal genomes (FAANG project, www.faang.org) started in 2014 [20] and will generate data that will foster understanding of how the genome is read and translated into complex animal traits of economic importance. Indeed, the recent explosion of data on the regulatory functions of ncRNAs proves their importance in the regulation of multiple/major biological processes impacting development, differentiation and metabolism. This chapter explores recent developments on the expression, regulation and functions of ncRNAs, in particular microRNA (miRNA) and long non-coding RNA (lncRNA), in ruminant (cattle, sheep and goat) mammary gland development and the lactation process, as well as illustrate our own studies on the roles of ncRNAs in these processes.

## 2. Non-coding RNAs: biosynthesis and classification

Non-coding RNAs are transcribed RNA molecules that are not translated into proteins. They play a remarkable variety of biological functions by engaging target transcripts through sequence-specific interactions. They regulate many biological processes, including gene expression (transcription, RNA processing and translation), protect genomes from foreign nucleic acids and can guide DNA synthesis or genome rearrangement [21]. In general, ncRNAs are classified according to size or function. According to size, ncRNAs are classified as (1) small or short ncRNA: <200 nucleotides in their mature forms (e.g. miRNA, PIWI-interacting RNA [piRNA], small nuclear RNA [snRNA], small nucleolar RNA [snoRNA] and endogenous small interfering RNA [siRNA]) and (2) long ncRNA: >200 nucleotides long (e.g. lncRNA). According to function, ncRNAs are classified as (1) housekeeping or translation-related ncRNAs: they are constitutively expressed and crucial for normal cellular function and viability and include tRNA, rRNA and snoRNA and (2) regulatory ncRNAs and include miRNA, lncRNA, siRNA and piRNA [22, 23]. The biogenesis of these various types of ncRNAs has been discussed extensively [23–26]. This chapter focuses particularly on the involvement of miRNA and lncRNA in ruminant mammary gland development and lactation.

### 2.1. MicroRNAs

MiRNAs are an abundant class of short ncRNAs of about 22 nucleotides long. They regulate a variety of cellular processes through post-transcriptional repression of gene expression. MiRNAs consequently control the activities of about 60% of all protein-coding genes and participate in the regulation of almost every cellular process investigated in mammals [25]. Mature miRNAs are generated from a series of biochemical events beginning in the nucleus and culminating in the cytoplasm [24, 27, 28]. Briefly, these events occur in several main steps as follows: (1) nuclear processing of primary miRNA transcripts (pri-miRNAs) into precursor miRNAs (pre-miRNAs) by the DiGeorge Syndrome Critical Region Gene 8 (DGCR8)/Drosha complex, (2) cytoplasmic processing of pre-miRNAs into imperfectly paired miRNA duplexes by dicer, and (3) preferential incorporation of one strand (the 'guide' miRNA strand) onto the RNA-induced silencing complex (RISC) [25]. Most miRNA genes located in introns of protein-coding genes share the promoter of the host gene [29]. MiRNAs often have multiple transcription start sites and regulate gene expression through inhibition of translation initiation or elongation, co-translational protein degradation and premature termination of translation [25, 30].

Since the discovery of the first miRNA, lin-4, in 1993 [31] and aided by deep sequencing technologies and developments in bioinformatics processing of deep sequence data, thousands of miRNAs have been detected in humans, mouse, farm animal species and plants and deposited in the miRNA data base (**Table 1**). Due to the crucial regulatory roles of miRNAs in many biological processes across species, they are being considered as candidate biomarkers of various human diseases, such as autoimmune [32], metabolic [33] and cardiovascular diseases [34], and various types of cancers [35–37].

| Species | MiRNA | | lncRNA | |
|---|---|---|---|---|
| | Precursor | Mature | Transcripts | Genes |
| Cattle | 808 | 793 | 22,386 | 23,696 |
| Sheep | 106 | 153 | – | |
| Goat | 267 | 436 | – | |
| Pig | 382 | 411 | – | |
| Chicken | 740 | 994 | 13,085 | 9681 |
| Human | 1881 | 2588 | 141,353 | 90,062 |
| Mouse | 1193 | 1915 | 117,405 | 79,940 |

*Data source: MiRBase release 21 (http://www.mirbase.org/[38], and NONCODE database (www.noncode.org, Noncode 2016 [39]).

**Table 1.** Number of detected miRNAs and lncRNAs in farm animal species, mouse and human*.

## 2.2. Long non-coding RNAs

Long non-coding RNAs are a diverse collection of non-coding RNAs with emerging regulatory roles in many biological processes in every branch of life [26, 40–42]. LncRNA transcripts are >200 nucleotides long and constitute the largest portion of the mammalian non-coding RNA transcriptome [40]. LncRNA closely resembles mRNA than other classes of ncRNA in terms of their biogenesis pathways and form. Most lncRNAs are transcribed by the activities of RNA polymerase II, have a 5′ terminal methylguanosine cap and are often spliced and polyadenylated [41]. Some non-polyadenylated lncRNAs arise through alternative pathways probably expressed from RNA polymerase III promoters [43, 44] or arise during splicing and small nucleolar RNA production [45]. Furthermore, some lncRNAs are regulated in different ways at different stages of their biogenesis, maturation and decay [26]. Thousands of genes encoding lncRNAs have been identified in mammalian genomes (including livestock species), birds and plants studied so far and deposited in the NONECODE database (www.nonecode.org [39], **Table 1**).

# 3. MicroRNA in mammary gland development and lactation biology

## 3.1. Occurrence of microRNA in ruminant mammary gland and in milk

The regulatory roles of miRNAs in livestock species have emerged and are growing quickly [46, 47]. The most recent release of miRBase (release 21, http://www.mirbase.org/, [38]) contains 793 mature miRNAs for cattle, 436 for goat and 153 for sheep [38] (**Table 1**). However, with the increase in the application of RNA sequencing in expression profiling of miRNAs in different livestock species, the number of novel livestock miRNAs is expected to increase.

### *3.1.1. Cattle*

The profiles of miRNAs in bovine MG tissue or milk have been investigated using different approaches, such as microarray [48, 49], genome sequencing [4] and RNA sequencing [50–57]. A total of 496 miRNA genes were identified following sequencing of the cattle genome of which 135 were novel [4].The expression profiles of miRNAs in MG tissues and cells facilitate discovery of novel miRNAs and also identification of candidate miRNAs for different cell types, lactation stages, periods, disease response and so on. Before the release of the bovine genome sequence, Gu et al. [49] pioneered miRNA discovery in the bovine MG by cloning and sequencing small RNAs from MG tissue followed by identification of 59 distinct bovine miRNAs. Using next-generation sequencing techniques, Chen et al. [58] identified 230 and 213 known miRNAs in cow colostrum and mature milk, respectively. The authors also observed that 108 and 8 miRNAs were upregulated and downregulated, respectively, in colostrum compared to mature milk [58]. Using microarray, Izumi et al. [59] identified 100 and 53 known miRNAs in colostrum and mature milk, respectively. Using Solexa sequencing method, Li et al. [60] reported 884 unique miRNAs sequences in the bovine MG (283 known, 505 novel and 96 conserved miRNAs). Le Guillou et al. [61] identify 167 novel miRNAs in the bovine MG, many of which were also detected in mouse MG. Analysing three milk fractions (fat, whey and cells) and mammary gland tissues, we reported 210, 200 and 249 known and 33, 31 and 36 novel miRNAs in milk fat, whey and cells, respectively, and 321 known and 176 novel miRNAs in mammary gland tissues [62]. Deep sequencing the milk fat across the lactation curve, we also identified a total of 475 known and 238 novel miRNAs [63].

### *3.1.2. Goat*

A total of 487 miRNAs were identified when the goat genome was sequenced and the largest miRNA clusters were found on chromosome 21 [6]. Using the Illumina-Solexa high-throughput sequencing technology to analyse goat MG tissues during early lactation, Ji et al. [64]

reported 131 novel and 300 conserved miRNAs. Using the same method (Illumina-Solexa sequencing), Li et al. [65] reported 346 conserved and 95 novel miRNAs in goat MG tissues from dry off and peak lactation does.

#### 3.1.3. *Sheep*

Most miRNAs identified in sheep come from tissues other than the MG. For example, Caiment et al. [66] identified 747 miRNAs from the skeletal muscle through deep sequencing, whereas McBride et al. [67] reported 212 miRNAs from sheep ovarian follicles and corpus lutea at various reproductive stages. In the MG, Galio et al. [68] showed the presence of three known miRNAs including miR-21, miR-205 and miR-200 family in pregnant and lactating sheep.

### 3.2. MicroRNA function in ruminant mammary gland and milk synthesis

#### 3.2.1. *Expression patterns of microRNAs in lactation stages*

##### 3.2.1.1. *Temporal and spatial expression of microRNAs*

Indication of involvement of miRNAs in MG functions was gained through observation of differences in type and expression levels of miRNAs between lactation stages, under different nutritional regimes and presence of disease pathogens. Li et al. [50] identified 56 miRNAs that were significantly differentially expressed between lactation and non-lactation periods. Similarly, Wang et al. [48] detected 12 downregulated miRNAs (miR-10a, miR-15b, miR-16, miR-21, miR-33b, miR-145, miR-146b, miR-155, miR-181a, miR-205, miR-221 and miR-223) in the dry period (30 days prepartum) compared to early lactation period (7 days postpartum) and one upregulated miRNA (miR-31) in early lactation compared to the dry period. Previously, we examined miRNA expression pattern during a lactation cycle to explore it regulatory mechanisms during lactation using milk fat as input tissue for sampling [63]. In a previous investigation, we have shown that milk fat miRNA transcriptome closely resemble the miRNome of MG tissue [62]. We collected samples at the lactogenesis (LAC) (day 1 and 7), galactopoiesis (GAL) (day 30, 70, 130, 170 and 230) and involution (INV) (day 290 and when milk production dropped to 5 kg/day) stages from nine cows for deep sequencing [63]. We observed that 15 miRNAs (miR-30a-5p, miR-30d, miR-21-5p, miR-26a, miR-148a, let-7a-5p, let-7b, let-7f, let-7g, miR-99a-5p, miR-191, miR-200a, miR-200c, miR-186, miR-92a) were highly expressed across lactation stages [63]. MiR-148a and miR-26a were the most abundantly expressed accounting for more than 10% of the read counts in each stage of lactation. We also performed a differential expression (DE) analysis and detected miR-29b/miR-363 and miR-874/miR-6254 as important mediators of transition signals from LAC to GAL and from GAL to INV stages, respectively [63]. Furthermore, DE analysis indicated various patterns of miRNA expression across the lactation curve. For instance, some miRNAs were highly expressed during early lactation (lactogenesis) followed by decreased expression at later stages, whereas others were slightly expressed during early lactation but showed increased expression during mid-lactation and decreased expression during late lactation and vice versa [63] (**Figure 2**).

(a)

| miRNAs/Stages[1] | GAL_vs_LAC | | INV_vs_LAC | | INV_vs_GAL | | Expression Pattern[2] |
|---|---|---|---|---|---|---|---|
| | Fold change | p_value | Fold change | p_value | Fold change | p_value | |
| miR-2285t | -2.69 | 1.99E-11 | -2.33 | 7.18E-24 | -6.28 | 9.32E-08 | D_D_D |
| miR-744 | -3.03 | 5.23E-13 | -2.22 | 7.54E-03 | 1.74 | 9.75E-04 | D_U_D |
| miR-221 | -2.66 | 3.49E-03 | 4.14 | 8.97E-04 | 11.00 | 3.67E-12 | D_U_U |
| miR-152 | 1.42 | 3.83E-02 | -1.60 | 4.84E-02 | -1.74 | 3.67E-06 | U_D_D |
| miR-29c | 6.82 | 2.34E-26 | 4.29 | 4.13E-10 | -1.57 | 1.89E-02 | U_D_U |
| miR-23a | 1.28 | 2.55E-02 | 2.22 | 4.90E-09 | 1.73 | 2.11E-06 | U_U_U |

(b)

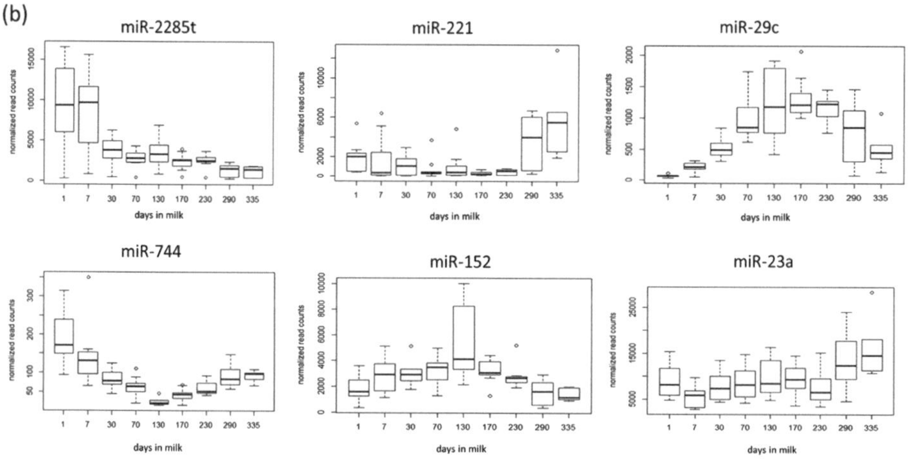


**Figure 2.** Differential miRNA expression patterns during a bovine lactation curve. (a) Fold change values of six miRNAs whose expression patterns changed significantly during each lactation switch and (b) box plots of their normalized read count values by lactation day. [1]LAC: lactogenesis; GAL: galactopoiesis; INV: involution; [2]D: downregulated and U: upregulated.

The temporal expression pattern of miRNAs has been reported in other ruminant species. For example, Galio et al. [68] reported a change in the expression pattern of miR-21, miR-205 and miR-200 family in MG tissues from pregnant and lactating sheep. From the early, middle and late stages of pregnancy and during lactation, the expression of miR-21 and miR-25 decreased, whereas miR-200 family (miR-200a, miR-200b, miR-200c, miR-141 and miR-429) showed increased expression [68]. Similarly, investigating the expression pattern of miRNAs during early and peak lactation and dry period, Li et al. [65] identified 15 differentially expressed miRNAs when comparing peak lactation and dry period including three significantly highly expressed miRNAs (miR-2887, miR-451 and miR-2478) during peak lactation and 12 significantly highly expressed miRNAs (miR-199b, miR-128, miR-25, miR-145, miR-98, miR-222, miR-181b, miR-199a-3p, miR-93, miR-221, let-7b and let-7c) during the dry period.

#### *3.2.1.2. MicroRNAs synergistically regulate lactation control mechanisms*

A wealth of evidence indicates that several miRNAs can work together to regulate target genes in the same or different biological pathways [69, 70]. We have successfully characterized a group of highly interacting miRNAs (modules) using a weighted co-expression network analysis [71] and correlated important miRNA modules to milk yield and milk components [72].

We identified three consensus (BLUE [62 miRNAs], TURQUOISE [133 miRNAs] and BROWN [59 miRNAs]) modules and the GREY module reserved for unclassified genes, throughout lactation stages (**Figure 3**). Based on module trait relationship, we were able to determine important modules (with absolute correlation >0.6) for milk components at each lactation stage. The BROWN and BLUE modules were highly related to protein and somatic cell count, respectively, in early lactation, the BLUE module to somatic cells in middle lactation and the BLUE module to urea and lactose in late lactation stage. We also found the most important component or hub miRNAs, which potentially coordinated miRNA synergetic mechanisms in their respective modules. MiR-149-5b and miR-874 were hub miRNAs in the BLUE module for milk somatic cells at early and middle lactation, respectively, whereas miR-330 was the hub miRNA in the BLUE module for milk urea and lactose at late lactation (**Figure 3**). Three miRNAs (mir-149-5b, miR-874 and miR-30) in the BLUE module play important roles in cell cycle [73–77], so it could be expected that these miRNAs regulate secretion of somatic cells in milk from MG.

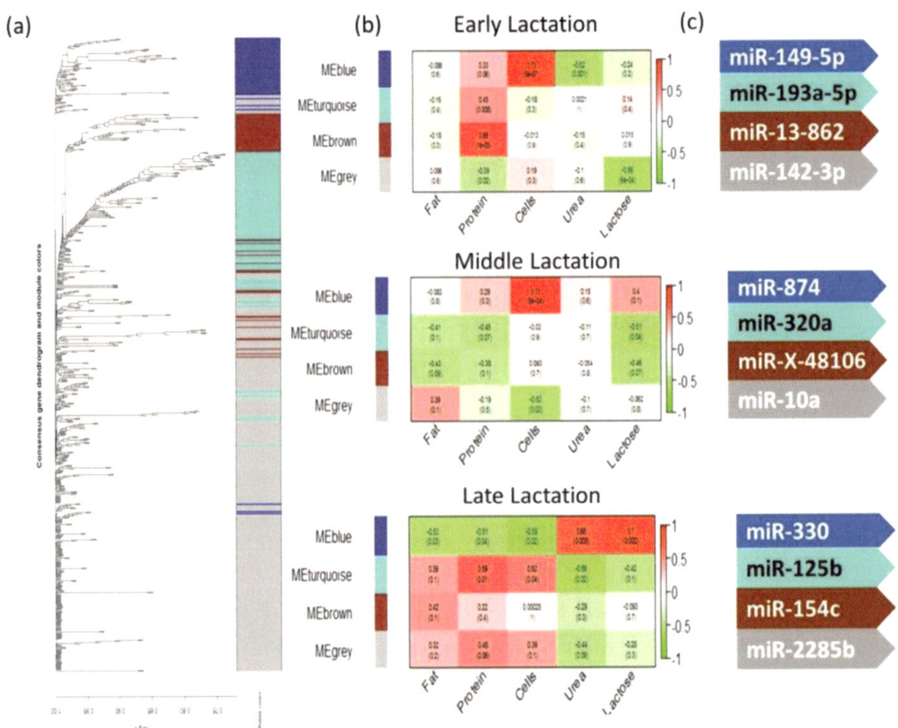


**Figure 3.** Important consensus modules and their hub miRNAs for milk component traits in different lactation periods. (a) Dynamic cut tree (dendrogram) based on topological overlap distance in gene expression profile; (b) module trait relationship in early, middle and late lactation and (c) hub miRNAs in the modules. GREY colour is for genes that do not belong to a specific module.

### *3.2.2. Networks and pathways regulated by microRNAs during a lactation cycle*

Through their target genes, miRNAs have been shown to control signal transduction in different species [78]. MiRNA roles in important pathways such as transforming growth factor beta

http://dx.doi.org/10.5772/67194

(TGF-β), prolactin and protein kinase signalling in MG development and lactation have been reviewed by several authors [79–83]. MiRNA regulation of three important signalling pathways (NOTCH, PTEN and HIPPO) in MG and breast cancer cells was recently reviewed [15]. Important miRNAs regulating these pathways include mir-34, mir-29, mir-146, mir-199 and mir-200 families for NOTCH signalling pathway, miR-21 and miR-155 for PTEN signalling pathway and miR-934 for HIPPO pathway. In Canadian Holstein cows, we performed the enrichment of differentially expressed miRNA target genes to signalling pathways and noted that relevant signalling pathways for transition between lactation stages are involved in apoptosis (PTEN and SAPK/JNK), intracellular signalling (protein kinase A, TGF-β and ERK5), cell cycle regulation (STAT3), cytokines (prolactin), hormone and growth factors (growth hormone and glucocorticoid receptor). *PTEN* is an important target gene for miR-29b in the regulation of mammary gland development [84]. PTEN signalling is crucial for the activities of prolactin autocrine [85]. The initiation of lactation is known to require induction of autocrine prolactin, and the level of this autocrine is known to be endogenously regulated by the signal of PTEN-PI3K-AKT pathway [85]. **Figure 4** is an illustration of some miRNAs that target genes in relevant signalling pathways during lactation [63]. Pathways, such as PTEN and growth hormone signalling, have been identified as important for regulatory mechanisms during lactation [85, 86].

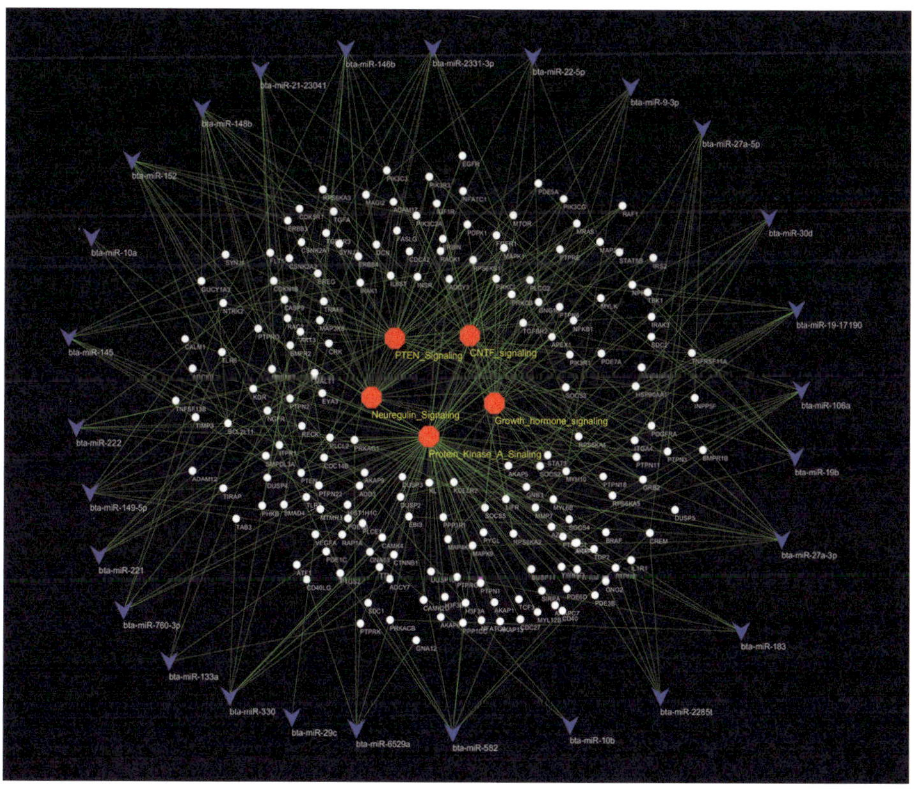


**Figure 4.** Illustration of miRNA-gene-pathway networks obtained from dynamic differentially expressed miRNAs during a bovine lactation curve. The outer layer shows miRNAs (blue arrow heads), which targets at least two genes (white dots) in significantly enriched pathways (red dots).

#### 3.2.3. *Functional validation of microRNA target genes*

Since *in vivo* experiments for functional validation of MG miRNAs are not feasible, such studies have mostly relied on the use of knock-out/mimics and MG-specific cell types. Using bovine mammary epithelial cells (BMEC), miR-15a was shown to regulate growth hormone receptor, viability of BMEC and the expression of casein genes [86]. MiR-486 regulation of lactation by targeting the PTEN gene in cow MGs has been demonstrated [87]. Bian et al. [88] recently reported that epigenetic regulation of miR-29s affects the lactation activity of BMEC. MiR-181a was shown to regulate the biosynthesis of bovine milk fat through targeting acyl-CoA synthetase long-chain family member 1 (*ACSL1*) [89]. MiR-103 was reported to control milk fat accumulation in goat MG during lactation [90]. Moreover, miR-27a was shown to suppress triglyceride accumulation as well as altered gene expression associated with fat metabolism in dairy goat mammary epithelial cells (GMEC) [91]. In another study, miR-135a was reported to target and regulate prolactin receptor (*PRLR*) gene in GMEC [92]. Inhibition of the expression of miR-145 in GMEC was shown to increase methylation levels of fatty acid synthase (*FASN*), stearoyl-CoA desaturase 1 (*SCD1*), peroxisome proliferator-activated receptor gamma (*PPARG*) and sterol regulatory element binding transcription factor 1 (*SREBF1*) [93]. MiR-24 control of triacylglycerol synthesis in goat mammary epithelial cells by targeting *FASN* gene has been demonstrated [94]. The ability of miR-145 to regulate lipogenesis in GMEC through targeting insulin-induced gene 1 (*INSIG1*) and epigenetic regulation of lipid-related genes has been demonstrated [93]. MiR-143 was shown to inhibit proliferation as well as induce apoptosis of GMEC [95]. MiR130b regulation of PPAR$\gamma$ coactivator-1$\alpha$ suppressed fat metabolism in GMEC [96]. In non-ruminant species, many miRNAs, including let-7 family members, mir-17/92, miR-30b, miR-93, miR-99a and miR-b, miR-101a, miR-126-3p, miR-138, miR-146b, miR-200 family members, mir-203, miR-205, miR-206, miR-210, miR-212/132, miR-221 and miR-424/50, have been reported to play roles in mammary gland development and disease [15]. Some miRNAs with functionally validated targets are summarized in **Table 2**.

### 3.3. Nutritional modulation of microRNA expression and function

The miRNA expression profile in response to dietary treatments has been studied in adipose tissues of lambs and cattle and bovine mammary gland tissues [56, 100–102]. A change in diet that interferes with energy balance has been shown to change miRNA expression pattern in cow liver [103]. Wang et al. [104] fed cows with high- and low-quality forage diets (corn stover and rice straw) and showed that miR-125b, miR-141, miR-181a, miR-221 and miR-15b changed their expression patterns across different tissues including MG. We have examined the expression pattern of miRNAs following MG adaptation to dietary supplementation with 5% linseed oil or 5% safflower oil using miRNA sequencing and identified seven differentially regulated miRNAs, including six upregulated (miR-199c, miR-199a-3p, miR-98, miR-378, miR-148b and miR-21-5p) and one downregulated (miR-200a) by both linseed and safflower oil. The target genes of these seven miRNAs have functions related to gene expression and general cellular metabolism and are enriched in four pathways of lipid metabolism (3-phosphoinositide biosynthesis, 3-phosphoinositide degradation, D-myo-inisitol-5-phosphate metabolism and the superpathway of inositol phosphate compounds) [51]. The largest number of target genes

(39) were associated with two functions (synthesis of lipid and concentration of lipid) related with lipogenesis. In goat, Mobuchon et al. [105] detected 30 miRNAs with expression patterns potentially modulated by food deprivation (14 and 16 were upregulated and downregulated, respectively). Among them, miR-204-5p and miR-223-3p were most remarkably affected by food deprivation and potentially played roles in the nutritional regulation of gene expression in the MG.

| MiRNAs | Target genes | Main consequence | Cell | References |
|---|---|---|---|---|
| miR-181 | *ACSL1* | Decrease lipid synthesis | BMEC | [89] |
| miR-29 family | *DNMT3A DNMT3B* | Decrease global DNA methylation | BMEC | [88] |
| miR-152 | *DNMT1* | Decrease global DNA methylation and increase expression of Akt and PPARγ | BMEC | [97] |
| miR-486 | *PTEN* | Alter expression of downstream genes of PTEN (AKT, mTOR pathways) | BMEC | [87] |
| miR-181b | *IRS2* | Wnt signalling pathway in GMEC | GMEC | [98] |
| miR-27a | *PPARγ* | Decrease triglyceride accumulation | GMEC | [91] |
| miR-26a and b | *INSIG1* | Decrease triacylglycerol synthesis | GMEC | [99] |
| miR-24 | *FASN, SREBF1, ACACA* | Decrease triacylglycerol synthesis | GMEC | [94] |
| miR-15a | *GHR* | Inhibit viability of mammary epithelial cells | BMEC | [86] |
| miR-130b | *PPARGC1A* | Repress PPARGC1A expression | GMEC | [96] |
| miR-143 | *BAX and BCL-2* | Inhibit proliferation and induce apoptosis | GMEC | [95] |
| miR145 | *INSIG1* | Increase fat droplet formation, triacylglycerol accumulation and proportion of unsaturated fatty acids | GMEC | [93] |

**Table 2.** MicroRNAs with functionally validated target genes using ruminant mammary gland cells.

### 3.4. MicroRNA functions in mammary gland health

MiRNAs have been shown to play roles in bovine infection and immunity in a wide range of tissues [54, 106–113]. For mammary gland, Naeem et al. [114] studied the expression of 14 miRNAs (miR-10a, miR-15b, miR-16a, miR-17, miR-21, miR-31, miR-145, miR-146a, miR-146b, miR-155, miR-181a, miR-205, miR-221 and miR-223) in MG tissue challenged with *Streptococcus uberis* and identified three downregulated miRNAs (miR 181a, miR 16 and miR-31) and one upregulated miRNA (miR-223) in infected versus healthy tissue. Lawless et al. [107] showed that 21 miRNAs were differentially expressed upon *Streptococcus uberis* infection of bovine primary epithelial cells. Using BMEC, Jin et al. [108] reported a differential expression of nine miRNAs (miR-184, miR-24-3p, miR-148, miR-486, let-7a-5p, miR-2339, miR-499, miR-23a and miR-99b) upon challenge with heat inactivated *Escherichia coli* and *Staphylococcus aureus* bacteria. Hou et al. [115] identified three upregulated miRNAs (miR-296, miR-2430 and miR-671) and one downregulated miRNA (miR-2318) in mastitis affected compared with healthy mammary gland tissues. Li et al. [111] sequenced RNA isolated from *S. aureus*-induced mastitis and control cows and identified 77 miRNAs with significant expression differences between the two groups. Li et al. [116] showed that miR-23 might be an important immune miRNA through its target mastitis candidate gene, high mobility group box 1 *(HMGB1)*.

### 3.5. MicroRNA function in milk recipients

Recent evidence suggesting that milk-derived miRNAs may have potential regulatory roles in modulating the immune system or metabolic processes of milk recipients still remain controversial [117–124]. Currently, there are two hypotheses about miRNA function in infants/offspring: the first proposes that milk miRNAs exert physiological regulatory functions after transferring to offspring, and the second assumes that miRNAs do not have any function but merely provide nutrition. According to Zhang et al. [117], the rice-derived miRNA, miR-168a, can bind to the mRNA of human/mouse low-density lipoprotein receptor adapter protein 1 (*LDLRAP1*) and inhibit its expression in the liver, and consequently decrease LDL removal from mouse plasma. Baier et al. [118] reported that miR-29b-3p and miR-200c-3p could be absorbed by humans in biologically meaningful amounts, which could affect related gene expression in peripheral blood mononuclear cells while Izumi et al. [125] confirmed that whey exosomes containing miRNAs and mRNA could be absorbed by human macrophages. These results opened a new aspect of the nutritional control of metabolism [119]. However, other studies have not succeeded to validate the hypothesis that milk miRNAs exert physiological regulatory functions after transferring to offspring [126–129]. For instance, Auerbach et al. [129] observed that drinking bovine milk increased circulating levels of miRNAs (miR-29b-3p and miR-200c-3p) but found no evidence that they significantly altered miRNA signals after milk ingestion. These authors concluded that milk miRNAs likely serve as a source of nutrition but not as post-transcriptional regulators in recipients.

## 4. Long non-coding RNA in mammary gland development and lactation biology

### 4.1. Prolife and expression of long non-coding RNAs

A limited number of studies have examined the occurrence and potential functions of lncRNAs in ruminant livestock species [130–132]. A pioneer study screened reconstructed transcript assemblies of bovine-specific expressed sequence tags and identified 449 putative lncRNAs located in 405 intergenic regions [130]. Following this initial study, Weikard et al. [131] used RNA sequencing technique and identified 4848 potential lncRNAs, which were predominantly intergenic (4365) in bovine skin. In another study, Billerey et al. [132] characterized 584 lncRNAs in bovine muscle in addition to significant correlated expression between 2083 pairs of lncRNA/protein encoding genes. Koufariotis et al. [133] characterized the lncRNA repertoire across 18 bovine tissues including the mammary gland and reported 9778 transcripts. Ibeagha-Awemu et al. [134] studied the lncRNA profile of the bovine mammary gland by RNA sequencing and identified 4227 lncRNAs (338 known and 3889 novel). In goats, Zhan et al. [135] sequenced libraries from developing longissimus dorsi fetal (45, 60 and 105 days of gestation) and postnatal (3 days after birth) muscles and identified 3981 lncRNA transcripts corresponding to 2739 lncRNA genes. Ren et al. [136] identified 1336 specific lncRNAs in fetal skin of Youzhou dark goat (dark skin) and Yudong white goat (white skin). Similarly, Chao et al. [137] in a study with aim to identify and classify new transcripts in Dorper and small-tail Han sheep muscle transcriptomes predicted with high confidence 1520 transcripts to be lncRNAs.

### 4.2. Function of long non-coding RNAs

While the regulatory roles of lncRNAs have been associated with several human disease conditions including tumourigenesis, cardiac development, aging and immune system development [138–143], little information exist on livestock species. Our previous study on bovine mammary gland identified 26 lncRNAs that were significantly differentially regulated in response to a diet rich in $\alpha$-linolenic acid thus suggesting potential regulatory roles of lncRNAs in fatty acid synthesis and lipid metabolism [134]. In a study with goat fetal muscle tissues at different stages of development, Zhan et al. [135] identified 577 significantly differentially expressed lncRNA transcripts thus suggesting roles in muscle development.

## 5. Genome editing technology and non-coding RNA

Genome engineering has been considered as the next genomic revolution [144], and it is expected to significantly improve livestock production by precision genome editing [145–147] favouring markers associated with improved productivity, reproduction and health status. The history of genome editing in livestock has been extensively reviewed [145, 148–150]. The advent of engineered endonucleases (EENs), including zinc finger nucleases (ZFNs) [151], transcription activator-like effector nucleases (TALENs) [152] and clustered regularly interspaced short palindromic repeats (CRISPR/Cas9) [153]), allows to cut a specific position in DNA sequence and then use endogenous cellular pathways to direct DNA repair to introduce specified alterations to the DNA sequence. Genome-editing approaches have been successfully used in different livestock species, such as pig [154, 155], goat [156], cattle [157] and sheep [158]. In dairy cows, these technologies have been used to manipulate the genome so that they produce specific milk types, such as milk that causes less allergic problems (e.g. milk with less β-lactoglobulin protein) [159, 160]. These genome-editing tools also helped to improve mammary gland health by generating mastitis-resistant cattle [161, 162]. From an animal breeding perspective, a simulation study showed that genomic prediction combined with genome editing could be of benefit [163]. A total of 10,000 additive loci were simulated and shown to contribute to the variation in selected traits and benefits could be achieved with only 20 of those loci being edited in each selected sire [163]. Similar to other genome sequences, miRNA gene sequences within mammalian genomes can be easily edited with high efficacy and precision [144]. Targeted miRNA editing will enable revelation of the complex regulatory circuits governed by miRNAs and realization, in the long term, of their full diagnostic and therapeutic potentials. For instance, Chen et al. [164] successfully used TALEN to disrupt the function of miR-21 in cancerous cells. A transgenic calf engineered to express miRNA-4 and miR-6 showed an absence of β-lactoglobulin and a concurrent increase in casein proteins in milk [165].

## 6. Conclusion and remarks

Up to now, it is well known that the mammalian genome encodes thousands of ncRNAs and these ncRNAs play important roles in many processes related to MG development, health and disease as well as roles in milk secretion and lactation processes. Regarding animal breeding, several ncRNAs target specific processes and their target genes could be important biomarkers for specific traits of interest. Therefore, the application of ncRNA to improve mammary gland health and milk production as well as enhance milk quality is very promising. However, the first step is a better understanding of ncRNA function in MG development and lactation. In fact, the MG is a complex tissue and lactation is a complicated process, but what we known about the regulatory networks underling MG function and the lactation process is very limited. For instance, through RNA sequencing, many novel ncRNAs have been detected in the MG but knowledge of their actual functions remains elusive. Therefore, integrated 'omics' approaches (genomics, transcriptomics, epigenomics and proteomics) should be used to identify and explore the potential roles of ncRNAs in mammary gland development and lactation biology. Moreover, a miRNA can target hundreds of genes thus making it difficult, costly and labour-intensive to functionally validate each miRNA gene target. Thus, integrative approaches such as combination of miRNA and mRNA expression in the same sample will refine computational predictions and increase our understanding of miRNA function and its application.

## Acknowledgements

We acknowledge financial support from Agriculture and Agri-Food Canada.

## Author details

Duy N. Do[1,2] and Eveline M. Ibeagha-Awemu[1*]

*Address all correspondence to: Eveline.ibeagha-awemu@agr.gc.ca

1 Agriculture and Agri-Food Canada, Sherbrooke Research and Development Centre, Sherbrooke, Quebec, Canada

2 Department of Animal Science, McGill University, Ste-Anne-De-Bellevue, Quebec, Canada

## References

[1] Lemay DG, Neville MC, Rudolph MC, Pollard KS, German JB: Gene regulatory networks in lactation: Identification of global principles using bioinformatics. BMC Systems Biology 2007, **1**(1):1-24.

[2] Strucken EM, Laurenson YCSM, Brockmann GA: Go with the flow—Biology and genetics of the lactation cycle. Frontiers in Genetics 2015, 6: 118 (11 pages).

[3] Zimin AV, Delcher AL, Florea L, Kelley DR, Schatz MC, Puiu D, Hanrahan F, Pertea G, Van Tassell CP, Sonstegard TS: A whole-genome assembly of the domestic cow, Bos taurus. Genome Biology 2009, **10**(4):1.

[4] Elsik CG, Tellam RL, Worley KC: The genome sequence of taurine cattle:a window to ruminant biology and evolution. Science 2009, **324**(5926):522-528.

[5] Jiang Y, Xie M, Chen W, Talbot R, Maddox JF, Faraut T, Wu C, Muzny DM, Li Y, Zhang W: The sheep genome illuminates biology of the rumen and lipid metabolism. Science 2014, **344**(6188):1168-1173.

[6] Dong Y, Xie M, Jiang Y, Xiao N, Du X, Zhang W, Tosser-Klopp G, Wang J, Yang S, Liang J: Sequencing and automated whole-genome optical mapping of the genome of a domestic goat (Capra hircus). Nature Biotechnology 2013, **31**(2):135-141.

[7] Matukumalli LK, Lawley CT, Schnabel RD, Taylor JF, Allan MF, Heaton MP, O'Connell J, Moore SS, Smith TP, Sonstegard TS: Development and characterization of a high density SNP genotyping assay for cattle. PLoS One 2009, **4**(4):e5350.

[8] Boichard D, Chung H, Dassonneville R, David X, Eggen A, Fritz S, Gietzen KJ, Hayes BJ, Lawley CT, Sonstegard TS: Design of a bovine low-density SNP array optimized for imputation. PLoS One 2012, **7**(3):e34130.

[9] Tosser-Klopp G, Bardou P, Bouchez O, Cabau C, Crooijmans R, Dong Y, Donnadieu-Tonon C, Eggen A, Heuven HC, Jamli S: Design and characterization of a 52K SNP chip for goats. PLoS One 2014, **9**(1):e86227.

[10] Archibald A, Cockett N, Dalrymple B, Faraut T, Kijas J, Maddox J, McEwan J, Hutton Oddy V, Raadsma H, Wade C: The sheep genome reference sequence: A work in progress. Animal Genetics 2010, **41**(5):449-453.

[11] Kijas JW, Townley D, Dalrymple BP, Heaton MP, Maddox JF, McGrath A, Wilson P, Ingersoll RG, McCulloch R, McWilliam S: A genome wide survey of SNP variation reveals the genetic structure of sheep breeds. PloS One 2009, **4**(3):e4668.

[12] Wickramasinghe S, Cánovas A, Rincón G, Medrano JF: RNA-sequencing: A tool to explore new frontiers in animal genetics. Livestock Science 2014, **166**:206-216.

[13] Osorio JS, Lohakare J, Bionaz M: Biosynthesis of milk fat, protein, and lactose: Roles of transcriptional and posttranscriptional regulation. Physiological Genomics 2016, **48**(4):231-256.

[14] Bionaz M, Loor J, Hurley W: Milk protein synthesis in the lactating mammary gland: Insights from transcriptomics analyses. INTECH Open Access Publisher, Rijeka, Croatia; 2012.

[15] Sandhu GK, Milevskiy MJ, Wilson W, Shewan AM, Brown MA: Non-coding RNAs in Mammary Gland Development and Disease. Advances in Experimental Medicine and Biology, Berlin, Germany, 2016, 886:121-153.

[16] Standaert L, Adriaens C, Radaelli E, Van Keymeulen A, Blanpain C, Hirose T, Nakagawa S, Marine J-C: The long noncoding RNA Neat1 is required for mammary gland development and lactation. RNA 2014, **20**(12):1844-1849.

[17] Hansji H, Leung EY, Baguley BC, Finlay GJ, Askarian-Amiri ME: Keeping abreast with long non-coding RNAs in mammary gland development and breast cancer. Frontiers in Genetics 2014, **5**:379.

[18] Birney E, Stamatoyannopoulos JA, Dutta A, Guigó R, Gingeras TR, Margulies EH, Weng Z, Snyder M, Dermitzakis ET, Thurman RE: Identification and analysis of functional elements in 1% of the human genome by the ENCODE pilot project. Nature 2007, **447**(7146):799-816.

[19] ENCODE_Project_Consortium: An integrated encyclopedia of DNA elements in the human genome. Nature 2012, **489**(7414):57-74.

[20] Andersson L, Archibald AL, Bottema CD, Brauning R, Burgess SC, Burt DW, Casas E, Cheng HH, Clarke L, Couldrey C: Coordinated international action to accelerate genome-to-phenome with FAANG, the Functional Annotation of Animal Genomes project. Genome Biology 2015, **16**(1):1.

[21] Cech TR, Steitz JA: The noncoding RNA revolution—trashing old rules to forge new ones. Cell 2014, **157**(1):77-94.

[22] Santosh B, Varshney A, Yadava PK: Non-coding RNAs: Biological functions and applications. Cell Biochemistry and Function 2015, **33**(1):14-22.

[23] Morris KV, Mattick JS: The rise of regulatory RNA. Nature Reviews Genetics 2014, **15**(6):423.

[24] Kim VN, Han J, Siomi MC: Biogenesis of small RNAs in animals. Nature Reviews Molecular Cell Biology 2009, **10**(2):126-139.

[25] Bartel DP: MicroRNAs. Cell 2004, **116**(2):281-297.

[26] Quinn JJ, Chang HY: Unique features of long non-coding RNA biogenesis and function. Nature Reviews Genetics 2016, **17**(1):47-62.

[27] Krol J, Loedige I, Filipowicz W: The widespread regulation of microRNA biogenesis, function and decay. Nature Reviews Genetics 2010, **11**(9):597-610.

[28] Filipowicz W, Bhattacharyya SN, Sonenberg N: Mechanisms of post-transcriptional regulation by microRNAs: are the answers in sight? Nature Reviews Genetics 2008, **9**(2):102-114.

[29] Borchert GM, Lanier W, Davidson BL: RNA polymerase III transcribes human microRNAs. Nature Structural & Molecular Biology 2006, **13**(12):1097-1101.

[30] Brodersen P, Voinnet O: Revisiting the principles of microRNA target recognition and mode of action. Nature reviews Molecular Cell Biology 2009, **10**(2):141-148.

[31] Lee RC, Feinbaum RL, Ambros V: The C. elegans heterochronic gene lin-4 encodes small RNAs with antisense complementarity to lin-14. Cell 1993, **75**(5):843-854.

[32] Pauley KM, Cha S, Chan EK: MicroRNA in autoimmunity and autoimmune diseases. Journal of Autoimmunity 2009, **32**(3):189-194.

[33] Rottiers V, Näär AM: MicroRNAs in metabolism and metabolic disorders. Nature Reviews Molecular Cell Biology 2012, **13**(4):239-250.

[34] Thum T: MicroRNA therapeutics in cardiovascular medicine. EMBO Molecular Medicine 2012, **4**(1):3-14.

[35] Alvarez-Garcia I, Miska EA: MicroRNA functions in animal development and human disease. Development 2005, **132**(21):4653-4662.

[36] Calin GA, Croce CM: MicroRNA signatures in human cancers. Nature Reviews Cancer 2006, **6**(11):857-866.

[37] Esquela-Kerscher A, Slack FJ: Oncomirs—microRNAs with a role in cancer. Nature Reviews Cancer 2006, **6**(4):259-269.

[38] Kozomara A, Griffiths-Jones S: miRBase: Annotating high confidence microRNAs using deep sequencing data. Nucleic Acids Research 2014, **42**(D1):D68-D73.

[39] Zhao Y, Li H, Fang S, Kang Y, Hao Y, Li Z, Bu D, Sun N, Zhang MQ, Chen R: NONCODE 2016: An informative and valuable data source of long non-coding RNAs. Nucleic Acids Research 2015:gkv1252.

[40] Mercer TR, Dinger ME, Mattick JS: Long non-coding RNAs: Insights into functions. Nature Reviews Genetics 2009, **10**(3):155-159.

[41] Mercer TR, Mattick JS: Structure and function of long noncoding RNAs in epigenetic regulation. Nature Structural & Molecular Biology 2013, **20**(3):300-307.

[42] Vance KW, Ponting CP: Transcriptional regulatory functions of nuclear long noncoding RNAs. Trends in Genetics 2014, **30**(8):348-355.

[43] Dieci G, Fiorino G, Castelnuovo M, Teichmann M, Pagano A: The expanding RNA polymerase III transcriptome. Trends in Genetics 2007, **23**(12):614-622.

[44] Kapranov P, Cheng J, Dike S, Nix DA, Duttagupta R, Willingham AT, Stadler PF, Hertel J, Hackermüller J, Hofacker IL: RNA maps reveal new RNA classes and a possible function for pervasive transcription. Science 2007, **316**(5830):1484-1488.

[45] Yin Q-F, Yang L, Zhang Y, Xiang J-F, Wu Y-W, Carmichael GG, Chen L-L: Long noncoding RNAs with snoRNA ends. Molecular Cell 2012, **48**(2):219-230.

[46] Wang X, Gu Z, Jiang H: MicroRNAs in farm animals. Animal 2013, **7**(10):1567-1575.

[47] Fatima A, Morris DG: MicroRNAs in domestic livestock. Physiological Genomics 2013, **45**(16):685-696.

[48] Wang M, Moisá S, Khan M, Wang J, Bu D, Loor J: MicroRNA expression patterns in the bovine mammary gland are affected by stage of lactation. Journal of Dairy Science 2012, **95**(11):6529-6535.

[49] Gu Z, Eleswarapu S, Jiang H: Identification and characterization of microRNAs from the bovine adipose tissue and mammary gland. FEBS Letters 2007, **581**(5):981-988.

[50] Li Z, Liu H, Jin X, Lo L, Liu J: Expression profiles of microRNAs from lactating and non-lactating bovine mammary glands and identification of miRNA related to lactation. BMC Genomics 2012, **13**(1):1.

[51] Wicik Z, Gajewska M, Majewska A, Walkiewicz D, Osińska E, Motyl T: Characterization of microRNA profile in mammary tissue of dairy and beef breed heifers. Journal of Animal Breeding and Genetics 2016, **133**(1):31-42.

[52] Wang D, Liang G, Wang B, Sun H, Liu J, Guan LL: Systematic microRNAome profiling reveals the roles of microRNAs in milk protein metabolism and quality: Insights on low-quality forage utilization. Scientific Reports 2016, **6**:21194.

[53] Li R, Dudemaine P-L, Zhao X, Lei C, Ibeagha-Awemu EM: Comparative analysis of the miRNome of bovine milk fat, whey and cells. PLoS One 2016, **11**(4):e0154129.

[54] Sun J, Aswath K, Schroeder SG, Lippolis JD, Reinhardt TA, Sonstegard TS: MicroRNA expression profiles of bovine milk exosomes in response to Staphylococcus aureus infection. BMC Genomics 2015, **16**(1):806.

[55] Li R, Zhang C-L, Liao X-X, Chen D, Wang W-Q, Zhu Y-H, Geng X-H, Ji D-J, Mao Y-J, Gong Y-C et al: Transcriptome microRNA profiling of bovine mammary glands infected with Staphylococcus aureus. International Journal of Molecular Sciences 2015, **16**(3):4997-5013.

[56] Li R, Beaudoin F, Ammah AA, Bissonnette N, Benchaar C, Zhao X, Lei C, Ibeagha-Awemu EM: Deep sequencing shows microRNA involvement in bovine mammary gland adaptation to diets supplemented with linseed oil or safflower oil. BMC Genomics 2015, **16**(1):1-16.

[57] Jin W, Ibeagha-Awemu EM, Liang G, Beaudoin F, Zhao X, Guan L: Transcriptome microRNA profiling of bovine mammary epithelial cells challenged with Escherichia

coli or Staphylococcus aureus bacteria reveals pathogen directed microRNA expression profiles. BMC Genomics 2014, 15:181 (16 pages).

[58] Chen X, Gao C, Li H, Huang L, Sun Q, Dong Y, Tian C, Gao S, Dong H, Guan D: Identification and characterization of microRNAs in raw milk during different periods of lactation, commercial fluid, and powdered milk products. Cell Research 2010, **20**(10):1128-1137.

[59] Izumi H, Kosaka N, Shimizu T, Sekine K, Ochiya T, Takase M: Bovine milk contains microRNA and messenger RNA that are stable under degradative conditions. Journal of Dairy Science 2012, **95**(9):4831-4841.

[60] Li Z, Liu H, Jin X, Lo L, Liu J: Expression profiles of microRNAs from lactating and non-lactating bovine mammary glands and identification of miRNA related to lactation. BMC Genomics 2012, **13**:731-731.

[61] Le Guillou S, Marthey S, Laloë D, Laubier J, Mobuchon L, Leroux C, Le Provost F: Characterisation and comparison of lactating mouse and bovine mammary gland miRNomes. PloS One 2014, **9**(3):e91938.

[62] Li R, Dudemaine PL, Zhao X, Lei C, Ibeagha-Awemu EM: Comparative analysis of the miRNome of bovine milk fat, whey and cells. PLoS One 2016, **11**.

[63] Do DN, Li R, Dudemaine P-L, Ibeagha-Awemu EM: MicroRNA roles in signaling during lactation: An insight from differential expression, time course and pathway analyses of deep sequence data. Scientific Reports 2017 (Revised version submitted).

[64] Ji Z, Wang G, Xie Z, Zhang C, Wang J: Identification and characterization of microRNA in the dairy goat (Capra hircus) mammary gland by Solexa deep-sequencing technology. Molecular Biology Reports 2012, **39**(10):9361-9371.

[65] Li Z, Lan X, Guo W, Sun J, Huang Y, Wang J, Huang T, Lei C, Fang X, Chen H: Comparative transcriptome profiling of dairy goat microRNAs from dry period and peak lactation mammary gland tissues. PLoS One 2012, **7**(12):e52388.

[66] Caiment F, Charlier C, Hadfield T, Cockett N, Georges M, Baurain D: Assessing the effect of the CLPG mutation on the microRNA catalog of skeletal muscle using high-throughput sequencing. Genome Research 2010, **20**(12):1651-1662.

[67] McBride D, Carre W, Sontakke S, Hogg CO, Law A, Donadeu FX, Clinton M: Identification of miRNAs associated with the follicular–luteal transition in the ruminant ovary. Reproduction 2012, **144**(2):221-233.

[68] Galio L, Droineau S, Yeboah P, Boudiaf H, Bouet S, Truchet S, Devinoy E: MicroRNA in the ovine mammary gland during early pregnancy: spatial and temporal expression of miR-21, miR-205, and miR-200. Physiological Genomics 2013, **45**(4):151-161.

[69] Yepes S, López R, Andrade RE, Rodriguez-Urrego PA, López-Kleine L, Torres MM: Co-expressed miRNAs in gastric adenocarcinoma. Genomics 2016, **108**(2):93-101.

[70] Wu J, Gong H, Bai Y, Zhang W: Analyzing the miRNA-gene networks to mine the important miRNAs under skin of human and mouse. BioMed Research International 2016: 5469371 (9 pages).

[71] Langfelder P, Horvath S: WGCNA: An R package for weighted correlation network analysis. BMC Bioinformatics 2008, 9:559 (13 pages).

[72] Ibeagha-Awemu E.M: MicroRNAs are master regulators of the bovine lactation curve. Invited presentation at Plant and Animal Genome XXIV Conference (Animal Epigenetics Workshop), January 9-13, 2016, San Diego, CA, USA

[73] Lee K-H, Chen Y-L, Yeh S-D, Hsiao M, Lin J, Goan Y, Lu P-J: MicroRNA-330 acts as tumor suppressor and induces apoptosis of prostate cancer cells through E2F1-mediated suppression of Akt phosphorylation. Oncogene 2009, **28**(38):3360-3370.

[74] Li Y, Zhu X, Xu W, Wang D, Yan J: miR-330 regulates the proliferation of colorectal cancer cells by targeting Cdc42. Biochemical and Biophysical Research Communications 2013, **431**(3):560-565.

[75] Wang L, Gao W, Hu F, Xu Z, Wang F: MicroRNA-874 inhibits cell proliferation and induces apoptosis in human breast cancer by targeting CDK9. FEBS Letters 2014, **588**(24):4527-4535.

[76] Lin RJ, Lin YC, Yu AL: miR-149* induces apoptosis by inhibiting Akt1 and E2F1 in human cancer cells. Molecular Carcinogenesis 2010, **49**(8):719-727.

[77] Pan S, Zhan S, Pei B, Sun Q, Bian L, Sun B: MicroRNA-149 inhibits proliferation and invasion of glioma cells via blockade of AKT1 signaling. International Journal of Immunopathology and Pharmacology 2012, **25**(4):871-881.

[78] Inui M, Martello G, Piccolo S: MicroRNA control of signal transduction. Nature Reviews Molecular Cell Biology 2010, **11**(4):252-263.

[79] Hennighausen L, Robinson GW, Wagner K-U, Liu X: Prolactin signaling in mammary gland development. Journal of Biological Chemistry 1997, **272**(12):7567-7569.

[80] Hennighausen L, Robinson GW: Signaling pathways in mammary gland development. Developmental Cell 2001, **1**(4):467-475.

[81] Hennighausen L, Robinson GW: Information networks in the mammary gland. Nature Reviews Molecular Cell Biology 2005, **6**(9):715-725.

[82] Callahan R, Raafat A: Notch signaling in mammary gland tumorigenesis. Journal of Mammary Gland Biology and Neoplasia 2001, **6**(1):23-36.

[83] Wakefield LM, Piek E, Böttinger EP: TGF-β signaling in mammary gland development and tumorigenesis. Journal of Mammary Gland Biology and Neoplasia 2001, **6**(1):67-82.

[84] Wang Z, Hou X, Qu B, Wang J, Gao X, Li Q: Pten regulates development and lactation in the mammary glands of dairy cows. PLoS One 2014, **9**(7):e102118.

[85] Chen C-C, Stairs DB, Boxer RB, Belka GK, Horseman ND, Alvarez JV, Chodosh LA: Autocrine prolactin induced by the Pten–Akt pathway is required for lactation initiation and provides a direct link between the Akt and Stat5 pathways. Genes & Development 2012, **26**(19):2154-2168.

[86] Li H-M, Wang C-M, Li Q-Z, Gao X-J: MiR-15a decreases bovine mammary epithelial cell viability and lactation and regulates growth hormone receptor expression. Molecules 2012, **17**(10):12037-12048.

[87] Li D, Xie X, Wang J, Bian Y, Li Q, Gao X, Wang C: MiR-486 regulates lactation and targets the PTEN gene in cow mammary glands. PLoS One 2015, **10**(3):e0118284.

[88] Bian Y, Lei Y, Wang C, Wang J, Wang L, Liu L, Liu L, Gao X, Li Q: Epigenetic regulation of miR-29s affects the lactation activity of dairy cow mammary epithelial cells. Journal of Cellular Physiology 2015, **230**(9):2152-2163.

[89] Lian S, Guo J, Nan X, Ma L, Loor J, Bu D: MicroRNA Bta-miR-181a regulates the biosynthesis of bovine milk fat by targeting ACSL1. Journal of Dairy Science 2016, **99**(5):3916-3924.

[90] Lin X, Luo J, Zhang L, Wang W, Gou D: MiR-103 controls milk fat accumulation in goat (Capra hircus) mammary gland during lactation. PLoS One 2013, **8**(11):e79258.

[91] Lin X-z, Luo J, Zhang L-p, Wang W, Shi H-b, Zhu J-j: MiR-27a suppresses triglyceride accumulation and affects gene mRNA expression associated with fat metabolism in dairy goat mammary gland epithelial cells. Gene 2013, **521**(1):15-23.

[92] Ji Z, Dong F, Wang G, Hou L, Liu Z, Chao T, Wang J: miR-135a targets and regulates prolactin receptor gene in goat mammary epithelial cells. DNA and Cell Biology 2015, **34**(8):534-540.

[93] Wang H, Shi H, Luo J, Yi Y, Yao D, Zhang X, Ma G, Loor JJ: MiR-145 regulates lipogenesis in goat mammary cells via targeting INSIG1 and epigenetic regulation of lipidrelated genes. Journal of Cellular Physiology 2016, 9999: 1–11.

[94] Wang H, Luo J, Chen Z, Cao W, Xu H, Gou D, Zhu J: MicroRNA-24 can control triacylglycerol synthesis in goat mammary epithelial cells by targeting the fatty acid synthase gene. Journal of Dairy Science 2015, **98**(12):9001-9014.

[95] Ji Z, Wang G, Hou L, Liu Z, Wang J, Chao T: miR-143 inhibits proliferation and induces apoptosis of mammary epithelial cells in dairy goat. Animal Cells and Systems 2016, **20**(2):63-69.

[96] Chen Z, Luo J, Ma L, Wang H, Cao W, Xu H, Zhu J, Sun Y, Li J, Yao D: MiR130b-regulation of PPARγ coactivator-1α suppresses fat metabolism in goat mammary epithelial cells. PLoS One 2015, **10**(11):e0142809.

[97] Wang J, Bian Y, Wang Z, Li D, Wang C, Li Q, Gao X: MicroRNA-152 regulates DNA methyltransferase 1 and is involved in the development and lactation of mammary glands in dairy cows. PLoS One 2014, **9**(7):e101358.

[98] Lian S, Guo JR, Nan XM, Ma L, Loor JJ, Bu DP: MicroRNA Bta-miR-181a regulates the biosynthesis of bovine milk fat by targeting ACSL1. Journal of Dairy Science 2016, **99**(5):3916-3924.

[99] Wang H, Luo J, Zhang T, Tian H, Ma Y, Xu H, Yao D, Loor JJ: MicroRNA-26a/b and their host genes synergistically regulate triacylglycerol synthesis by targeting the INSIG1 gene. RNA Biology 2016:1-11.

[100] Romao JM, Jin W, He M, McAllister T, Guan LL: Altered microRNA expression in bovine subcutaneous and visceral adipose tissues from cattle under different diet. PLoS One 2012, 7:e40605 (10 pages).

[101] Meale S, Romao J, He M, Chaves A, McAllister T, Guan L: Effect of diet on microRNA expression in ovine subcutaneous and visceral adipose tissues. Journal of Animal Science 2014, **92**(8):3328-3337.

[102] Romao J, Jin W, He M, McAllister T, Guan L: MicroRNAs in bovine adipogenesis: Genomic context, expression and function. BMC Genomics 2014, 15:137 (15 pages).

[103] Fatima A, Waters S, O'Boyle P, Seoighe C, Morris DG: Alterations in hepatic miRNA expression during negative energy balance in postpartum dairy cattle. BMC Genomics 2014, **15**(1):1.

[104] Wang D, Liang G, Wang B, Sun H, Liu J, Guan LL: Systematic microRNAome profiling reveals the roles of microRNAs in milk protein metabolism and quality: Insights on low-quality forage utilization. Scientific Reports 2016, 6: 21194 (16 pages).

[105] Mobuchon L, Marthey S, Le Guillou S, Laloë D, Le Provost F, Leroux C: Food deprivation affects the miRNome in the lactating goat mammary gland. PLoS One 2015, **10**(10):e0140111.

[106] Lawless N, Vegh P, O'Farrelly C, Lynn DJ: The Role of microRNAs in Bovine Infection and Immunity. Frontiers in Immunology 2014, 5:611 (7 pages).

[107] Lawless N, Foroushani AB, McCabe MS, O'Farrelly C, Lynn DJ: Next generation sequencing reveals the expression of a unique miRNA profile in response to a gram-positive bacterial infection. PLoS One 2013, **8**(3):e57543.

[108] Jin W, Ibeagha-Awemu EM, Liang G, Beaudoin F, Zhao X: Transcriptome microRNA profiling of bovine mammary epithelial cells challenged with Escherichia coli or Staphylococcus aureus bacteria reveals pathogen directed microRNA expression profiles. BMC Genomics 2014, 15: 181 (16 pages).

[109] Vegh P, Foroushani AB, Magee DA, McCabe MS, Browne JA, Nalpas NC, Conlon KM, Gordon SV, Bradley DG, MacHugh DE: Profiling microRNA expression in bovine alveolar macrophages using RNA-seq. Veterinary Immunology And Immunopathology 2013, **155**(4):238-244.

[110] Lawless N, Reinhardt TA, Bryan K, Baker M, Pesch B, Zimmerman D, Zuelke K, Sonstegard T, O'Farrelly C, Lippolis JD: MicroRNA regulation of bovine monocyte inflammatory and metabolic networks in an in vivo infection model. G3: Genes| Genomes| Genetics 2014, **4**(6):957-971.

[111] Li R, Zhang C-L, Liao X-X, Chen D, Wang W-Q, Zhu Y-H, Geng X-H, Ji D-J, Mao Y-J, Gong Y-C: Transcriptome microRNA profiling of bovine mammary glands infected with Staphylococcus aureus. International Journal of Molecular Sciences 2015, **16**(3):4997-5013.

[112] Liang G, Malmuthuge N, Griebel P: Model systems to analyze the role of miRNAs and commensal microflora in bovine mucosal immune system development. Molecular Immunology 2015, **66**(1):57-67.

[113] Salilew-Wondim D, Ibrahim S, Gebremedhn S, Tesfaye D, Heppelmann M, Bollwein H, Pfarrer C, Tholen E, Neuhoff C, Schellander K: Clinical and subclinical endometritis induced alterations in bovine endometrial transcriptome and miRNome profile. BMC Genomics 2016, **17**(1):1.

[114] Naeem A, Zhong K, Moisá S, Drackley J, Moyes K, Loor J: Bioinformatics analysis of microRNA and putative target genes in bovine mammary tissue infected with Streptococcus uberis. Journal of Dairy Science 2012, **95**(11):6397-6408.

[115] Hou Q, Huang J, Ju Z, Li Q, Li L, Wang C, Sun T, Wang L, Hou M, Hang S: Identification of splice variants, targeted microRNAs and functional single nucleotide polymorphisms of the BOLA-DQA2 gene in dairy cattle. DNA and Cell Biology 2012, **31**(5):739-744.

[116] Li L, Huang J, Zhang X, Ju Z, Qi C, Zhang Y, Li Q, Wang C, Miao W, Zhong J: One SNP in the 3′-UTR of HMGB1 gene affects the binding of target bta-miR-223 and is involved in mastitis in dairy cattle. Immunogenetics 2012, **64**(11):817-824.

[117] Zhang L, Hou D, Chen X, Li D, Zhu L, Zhang Y, Li J, Bian Z, Liang X, Cai X: Exogenous plant MIR168a specifically targets mammalian LDLRAP1: Evidence of cross-kingdom regulation by microRNA. Cell Research 2012, **22**(1):107-126.

[118] Baier SR, Nguyen C, Xie F, Wood JR, Zempleni J: MicroRNAs are absorbed in biologically meaningful amounts from nutritionally relevant doses of cow milk and affect gene expression in peripheral blood mononuclear cells, HEK-293 kidney cell cultures, and mouse livers. The Journal of Nutrition 2014, **144**(10):1495-1500.

[119] Liang H, Huang L, Cao J, Zen K, Chen X, Zhang CY: Regulation of mammalian gene expression by exogenous microRNAs. Wiley Interdisciplinary Reviews: RNA 2012, **3**(5):733-742.

[120] Denzler R, Stoffel M: Uptake and function studies of maternal milk-derived microRNAs. Journal of Biological Chemistry 2015, **290**(39):23680-23691.

[121] Van Rooij E, Bushell M, Zhang CY, Dashwood R, James W, Harris C, Baltimore D: The role of microRNA in nutritional control. Journal of Internal Medicine 2015, **278**(2):99-109.

[122] Fabris L, Calin GA: Circulating free xeno-microRNAs—The new kids on the block. Molecular Oncology 2016, **10**(3):503-508.

[123] Hirschi KD, Pruss GJ, Vance V: Dietary delivery: A new avenue for microRNA therapeutics? Trends in Biotechnology 2015, **33**(8):431-432.

[124] Wolf T, Baier SR, Zempleni J: The intestinal transport of bovine milk exosomes is mediated by endocytosis in human colon carcinoma Caco-2 cells and rat small intestinal IEC-6 cells. The Journal of Nutrition 2015, **145**(10):2201-2206.

[125] Izumi H, Tsuda M, Sato Y, Kosaka N, Ochiya T, Iwamoto H, Namba K, Takeda Y: Bovine milk exosomes contain microRNA and mRNA and are taken up by human macrophages. Journal of Dairy Science 2015, **98**(5):2920-2933.

[126] Micó V, Martín R, Lasunción MA, Ordovás JM, Daimiel L: Unsuccessful detection of plant microRNAs in beer, extra virgin olive oil and human plasma after an acute ingestion of extra virgin olive oil. Plant Foods for Human Nutrition 2016, **71**(1):102-108.

[127] Melnik BC, Kakulas F, Geddes DT, Hartmann PE, John SM, Carrera-Bastos P, Cordain L, Schmitz G: Milk miRNAs: Simple nutrients or systemic functional regulators? Nutrition & Metabolism 2016, **13**(1):1.

[128] Masood M, Everett CP, Chan SY, Snow JW: Negligible uptake and transfer of diet-derived pollen microRNAs in adult honey bees. RNA biology 2016, **13**(1):109-118.

[129] Auerbach A, Vyas G, Li A, Halushka M, Witwer K: Uptake of dietary milk miRNAs by adult humans: A validation study. F1000Research 2016, **5**:721.

[130] Huang W, Long N, Khatib H: Genome-wide identification and initial characterization of bovine long non-coding RNAs from EST data. Animal Genetics 2012, **43**(6):674-682.

[131] Weikard R, Hadlich F, Kuehn C: Identification of novel transcripts and noncoding RNAs in bovine skin by deep next generation sequencing. BMC Genomics 2013, 14:789 (15 pages).

[132] Billerey C, Boussaha M, Esquerré D, Rebours E, Djari A, Meersseman C, Klopp C, Gautheret D, Rocha D: Identification of large intergenic non-coding RNAs in bovine muscle using next-generation transcriptomic sequencing. BMC Genomics 2014, 15:499 (10 pages).

[133] Koufariotis LT, Chen Y-PP, Chamberlain A, Vander Jagt C, Hayes BJ: A catalogue of novel bovine long noncoding RNA across 18 tissues. PLoS One 2015, **10**(10):e0141225.

[134] Ibeagha-Awemu EM, Li R, Dudemaine P-L: The long non-coding RNA transcriptome of the bovine mammary gland and potential regulatory roles in fatty acid synthesis. Proceedings of the 6th International Symposium on Animal Functional Genomics

(6th ISFAG) 2015, Piacenza, 91. Available at: http://www.isafg2015.it/ISAFG2015_PROCEEDINGS.pdf.

[135] Zhan S, Dong Y, Zhao W, Guo J, Zhong T, Wang L, Li L, Zhang H: Genome-wide identification and characterization of long non-coding RNAs in developmental skeletal muscle of fetal goat. BMC Genomics 2016, **17**(1):666.

[136] Ren H, Wang G, Chen L, Jiang J, Liu L, Li N, Zhao J, Sun X, Zhou P: Genome-wide analysis of long non-coding RNAs at early stage of skin pigmentation in goats (Capra hircus). BMC Genomics 2016, **17**(1):67.

[137] Chao T, Wang G, Wang J, Liu Z, Ji Z, Hou L, Zhang C: Identification and classification of new transcripts in Dorper and small-tailed Han sheep skeletal muscle transcriptomes. PLoS One 2016, **11**(7):e0159638.

[138] Wapinski O, Chang HY: Long noncoding RNAs and human disease. Trends in Cell Biology 2011, **21**(6):354-361.

[139] Cheetham S, Gruhl F, Mattick J, Dinger M: Long noncoding RNAs and the genetics of cancer. British Journal of Cancer 2013, **108**(12):2419-2425.

[140] Esteller M: Non-coding RNAs in human disease. Nature Reviews Genetics 2011, **12**(12):861-874.

[141] Atianand MK, Fitzgerald KA: Long non-coding RNAs and control of gene expression in the immune system. Trends in Molecular Medicine 2014, **20**(11):623-631.

[142] Devaux Y, Zangrando J, Schroen B, Creemers EE, Pedrazzini T, Chang C-P, Dorn II GW, Thum T, Heymans S: Long noncoding RNAs in cardiac development and ageing. Nature Reviews Cardiology 2015, **12**(7):415-425.

[143] Kim J, Kim KM, Noh JH, Yoon J-H, Abdelmohsen K, Gorospe M: Long noncoding RNAs in diseases of aging. Biochimica et Biophysica Acta (BBA) - Gene Regulatory Mechanisms 2016, **1859**(1):209-221.

[144] Yu L, Batara J, Lu B: Application of genome editing technology to microRNA research in mammalians; 2016. Modern Tools for Genetic Engineering, InTech, Rijeka, Croatia, DOI: 10.5772/64330.

[145] Laible G, Wei J, Wagner S: Improving livestock for agriculture–technological progress from random transgenesis to precision genome editing heralds a new era. Biotechnology Journal 2015, **10**(1):109-120.

[146] West J, Gill WW: Genome editing in large animals. Journal of Equine Veterinary Science 2016, **41**:1-6.

[147] Tizard M, Hallerman E, Fahrenkrug S, Newell-McGloughlin M, Gibson J, de Loos F, Wagner S, Laible G, Han JY, D'Occio M: Strategies to enable the adoption of animal biotechnology to sustainably improve global food safety and security. Transgenic Research 2016:1-21.

[148] Proudfoot C, Carlson DF, Huddart R, Long CR, Pryor JH, King TJ, Lillico SG, Mileham AJ, McLaren DG, Whitelaw CBA: Genome edited sheep and cattle. Transgenic Research 2015, **24**(1):147-153.

[149] Carlson DF, Walton MW, Fahrenkrug SC, Hackett PB: Precision editing of large animal genomes. Advances in Genetics 2012, **80**:37.

[150] Butler JR, Ladowski JM, Martens GR, Tector M, Tector AJ: Recent advances in genome editing and creation of genetically modified pigs. International Journal of Surgery 2015, **23**:217-222.

[151] Kim Y-G, Cha J, Chandrasegaran S: Hybrid restriction enzymes: Zinc finger fusions to Fok I cleavage domain. Proceedings of the National Academy of Sciences 1996, **93**(3):1156-1160.

[152] Boch J, Scholze H, Schornack S, Landgraf A, Hahn S, Kay S, Lahaye T, Nickstadt A, Bonas U: Breaking the code of DNA binding specificity of TAL-type III effectors. Science 2009, **326**(5959):1509-1512.

[153] Jinek M, Chylinski K, Fonfara I, Hauer M, Doudna JA, Charpentier E: A programmable dual-RNA–guided DNA endonuclease in adaptive bacterial immunity. Science 2012, **337**(6096):816-821.

[154] Lillico SG, Proudfoot C, Carlson DF, Stverakova D, Neil C, Blain C, King TJ, Ritchie WA, Tan W, Mileham AJ: Live pigs produced from genome edited zygotes. Scientific Reports 2013, **3**:2847.

[155] Whitworth KM, Lee K, Benne JA, Beaton BP, Spate LD, Murphy SL, Samuel MS, Mao J, O'Gorman C, Walters EM: Use of the CRISPR/Cas9 system to produce genetically engineered pigs from in vitro-derived oocytes and embryos. Biology of Reproduction 2014, **91**(3):78.

[156] Ge H, Cui C, Liu J, Luo Y, Quan F, Jin Y, Zhang Y: The growth and reproduction performance of TALEN-mediated β-lactoglobulin-knockout bucks. Transgenic Research 2016:1-9.

[157] Wu H, Wang Y, Zhang Y, Yang M, Lv J, Liu J, Zhang Y: TALE nickase-mediated SP110 knockin endows cattle with increased resistance to tuberculosis. Proceedings of the National Academy of Sciences 2015, **112**(13):E1530-E1539.

[158] Zhang C, Wang L, Ren G, Li Z, Ren C, Zhang T, Xu K, Zhang Z: Targeted disruption of the sheep MSTN gene by engineered zinc-finger nucleases. Molecular Biology Reports 2014, **41**(1):209-215.

[159] Cui C, Song Y, Liu J, Ge H, Li Q, Huang H, Hu L, Zhu H, Jin Y, Zhang Y: Gene targeting by TALEN-induced homologous recombination in goats directs production of β-lactoglobulin-free, high-human lactoferrin milk. Scientific Reports 2015, **5**:10482.

[160] Yu S, Luo J, Song Z, Ding F, Dai Y, Li N: Highly efficient modification of beta-lactoglobulin (BLG) gene via zinc-finger nucleases in cattle. Cell Research 2011, **21**(11):1638.